LOCAL PLANNING FOR
TERROR AND DISASTER

LOCAL PLANNING FOR TERROR AND DISASTER

From Bioterrorism to Earthquakes

Edited by

LEONARD A. COLE
NANCY D. CONNELL

WILEY-BLACKWELL

A John Wiley & Sons, Inc., Publication

Published by John Wiley & Sons, Inc., Hoboken, New Jersey.
Published simultaneously in Canada.

For general information on our other products and services or for technical support, please contact our Customer Care Department within the United States at (800) 762-2974, outside the United States at (317) 572-3993 or fax (317) 572-4002.

Wiley also publishes its books in a variety of electronic formats. Some content that appears in print may not be available in electronic formats. For more information about Wiley products, visit our web site at www.wiley.com.

Library of Congress Cataloging-in-Publication Data

Local planning for terror and disaster : from bioterrorism to earthquakes / Leonard A. Cole, Nancy D. Connell, editors.
    p. cm.
  Includes index.
  ISBN 978-1-118-11286-1 (pbk.)
  1. Emergency management–Case studies. 2. Disaster relief–Case studies. 3. Disaster medicine–Case studies. I. Cole, Leonard A., 1933– II. Connell, Nancy D., 1952–
  HV551.2.L63 2012
  363.34′561–dc23

                                               2012015250

Printed in the United States of America.

10 9 8 7 6 5 4 3 2 1

CONTENTS

CONTRIBUTORS

Bruria Adini, PhD, is Senior Consultant to the Israeli Ministry of Health and a member of the Department of Emergency Medicine at Ben-Gurion University of the Negev. An expert on emergency disaster training, during her service in the Israel Defense Forces she headed the Medical Corps' emergency hospitalization branch, which was responsible for emergency preparedness of all general hospitals.

Limor Aharonson-Daniel, PhD, heads the Department of Emergency Medicine and is the founding director of the PREPARED Center for Emergency Response Research at Ben-Gurion University of the Negev in Israel. She has published extensively on injury research methods and on disaster preparedness assessment.

Isaac Ashkenazi, MD, MSc, MPA, MNS, is Director of Urban Terrorism Preparedness at the National Preparedness Leadership Institute, Harvard University. He is also Professor of Disaster Medicine and an international expert on Crisis Management and Leadership at Ben Gurion University of the Negev in Israel and is former Surgeon General for the Israel Defense Forces Home Front Command.

Steven M. Becker, PhD, is Professor of Community and Environmental Health, College of Health Sciences, at Old Dominion University in Virginia. A leading authority on the public health and risk communication aspects of disasters, emergencies, and terrorism, he was a member of a Radiological Emergency Assistance Mission invited to Japan after the March 2011 earthquake-tsunami disaster and the Fukushima Dai-ichi nuclear accident.

Lisa M. Brown, PhD, is a licensed clinical psychologist and Associate Professor in the School of Aging Studies, College of Behavioral and Community Sciences, University of South Florida. She is especially knowledgeable about challenges to non-profit organizations and government agencies during catastrophic events, and her research focus is on disaster planning, response, and recovery for individuals and communities.

Ronald V. Clarke, PhD, University Professor at Rutgers and a Visiting Professor at University College, London, worked for 15 years in the British government's criminological research department before moving to the United States. He is the founding editor of *Crime Prevention Studies* and his publications include *Superhighway Robbery: Preventing E-commerce Crime* (Willan Publishing, 2003) and *Outsmarting the Terrorists* (Praeger, 2006), both with Graeme Newman.

Leonard A. Cole, PhD, DDS, is an Adjunct Professor of Political Science at Rutgers University–Newark and Director of the Program on Terror Medicine and Security at the UMDNJ Center for BioDefense. His numerous publications on terrorism-related subjects include *The Anthrax Letters* (revised edition, Skyhorse, 2009) and *Essentials of Terror Medicine* (co-editor, Springer, 2009).

Nancy D. Connell, PhD, is Professor of Infectious Disease at the Medical School of the University of Medicine and Dentistry of New Jersey. She is Director of the UMDNJ Center for BioDefense, has authored many scientific articles, and has served on several panels concerning biosecurity, including the National Academy of Sciences committee that reviewed the FBI's scientific investigation of the 2001 anthrax mailings.

Henry P. Cortacans, MA, a Certified Emergency Manager and Nationally Registered Paramedic, is the State Planner for the New Jersey EMS Task Force. He was a responder to the 9/11 attacks and the US Airways Flight 1549 Miracle on the Hudson and is a recipient of the Interagency Disaster Preparedness Award Certificate of Merit from the International Association of Emergency Managers.

Steven M. Crimando, MA, BCETS, CTS, CHS-V, is an expert on the application of the behavioral sciences in homeland security, violence prevention, and disaster management. His focus is on the emotional and behavioral aspects associated with mass violence, including chemical, biological, and radiological threats, active shooter response, and collective violence related to group and crowd behavior.

Dian Dowling Evans, PhD, FNP-BC, is an Assistant Clinical Professor in the Nell Hodgson Woodruff School of Nursing at Emory University where she directs the graduate Emergency Nurse Practitioner specialty program. She has been practicing clinically as an emergency nurse practitioner since 1990 and has published on a wide range of emergency and advanced practice nursing topics.

Henry Falk, MD, MPH, consultant to the Office of Noncommunicable Disease, Injury and Environmental Health at the Centers for Disease Control and Prevention, is an Adjunct Professor of Environmental Health at the Emory University Rollins School of Public Health. Formerly he was the Director of the National Center for Environmental Health at the CDC, and of the Agency for Toxic Substances and Disease Registry.

David L. Glotzer, DDS, is a Clinical Professor at the Department of Cariology and Comprehensive Care, New York University College of Dentistry (NYUCD). He is a former Colonel in the US Army, and since 9/11 he has published on and helped lead an NYUCD initiative to define a role for dentists in public health response to major terrorist and disaster situations.

James W. Gordon, MPAS, PA-C, is Director of Physician Assistant/Nurse Practitioner Education at Emory University School of Medicine. He is Medical Officer for a Disaster Medical Assistance Team (part of the National Disaster Medical System), was a responder at the 1996 Olympic Park bombing, and was the medical point man on two Mount Everest expeditions.

Gerard A. Jacobs, PhD, is Director of the Disaster Mental Health Institute and a Professor in the Clinical Psychology Training Program at the University of South Dakota. He has helped develop disaster mental health and psychological support programs (such as community-based psychological first aid) throughout the world and was the recipient in 2007 of the American Psychological Association's International Humanitarian Award.

Donald H. Jenkins, MD, FACS, Associate Professor of Surgery and Director of Trauma at the Mayo Clinic in Rochester, Minnesota, is Vice-Chair of the National Trauma Institute. A former Colonel in the US Air Force, he served as Medical Director of the Trauma System for the US Central Command (including Afghanistan and Iraq).

Laura H. Kahn, MD, MPH, MPP, a general internist, is a research scholar with the Program on Science and Global Security at the Woodrow Wilson School of Public and International Affairs, Princeton University. She is the author of *Who's in Charge? Leadership During Epidemics, Bioterror Attacks, and other Public Health Crises* (Praeger Security International, 2009), is a monthly columnist for the *Bulletin of the Atomic Scientists*, and is co-founder of the One Health Initiative.

Emily G. Kidd, MD, is an Assistant Professor at the University of Texas Health Science Center at San Antonio and the Assistant Medical Director for the San Antonio Fire Department. She is also the Project Director for the Texas Disaster Medical System, a statewide collaboration of public health and acute medical care initiatives to improve disaster response across the state of Texas.

Craig A. Manifold, DO, is an Assistant Professor at the University of Texas Health Science Center at San Antonio and the Medical Director for the San Antonio Fire Department. He has deployed with the US Air Force and Texas Air National Guard in support of numerous large-scale medical operations and was the lead physician for relief operations during Hurricanes Katrina, Rita, Gustav, and Ike.

Brendan McCluskey, JD, MPA, CEM, is Executive Director of Emergency Management and Occupational Health and Safety at the University of Medicine and Dentistry of New Jersey. He previously served as an Emergency Medical Technician and paramedic and as Deputy Director, UMDNJ Center for BioDefense at the New Jersey Medical School.

Mark A. Merlin, DO, an emergency physician and Adjunct Associate Professor at the University of Medicine and Dentistry of New Jersey, is the Medical Director of the New Jersey EMS Task Force. He is also EMS/Disaster Medicine Fellowship Director at Newark Beth Israel Medical Center/Barnabas Health in Newark, New Jersey, and has published numerous papers on prehospital care.

Njoki Mwarumba, PhD candidate at Oklahoma State University, earned a BA in Communications and Community Development in Kenya. Her doctoral studies are in OSU's Fire and Emergency Management Program with a focus on complex emergencies in least developed countries.

James Netterwald, PhD, a medical technologist accredited by the American Society for Clinical Pathology, is President and CEO of BioPharmaComm, LLC. He has published articles on issues ranging from genomics and biotechnologies to drug discovery and development process.

Graeme R. Newman, PhD, is distinguished teaching professor at the School of Criminal Justice, University at Albany, and Associate Director of the Center for Problem-Oriented Policing. His publications include *The Punishment Response* (Transaction, 2008), *Comparative Deviance* (Transaction, 2007) and, with Ronald Clarke, *Superhighway Robbery: Preventing E-commerce Crime* (Willan Publishing, 2003) and *Outsmarting the Terrorists* (Praeger, 2006).

Ann E. Norwood, MD, Senior Associate at the Center for Biosecurity of UPMC, is a psychiatrist who retired from the US Army after 26 years of service. A former Chair of the American Psychiatric Association's Committee on Psychiatric Dimensions of Disaster, she has co-edited four books and published numerous articles and chapters on the behavioral health aspects of trauma associated with war, terrorism, and disasters.

Brenda D. Phillips, PhD, is Professor of Emergency Management at Oklahoma State University. She is also a Senior Researcher with the Center for the Study of Disasters and Extreme Events at OSU where she specializes in socially vulnerable populations, volunteer management, and long-term community recovery from disasters.

Peter M. Sandman, PhD, is a self-employed risk communication consultant based in Princeton, New Jersey. He has advised many clients on pre-crisis, mid-crisis, and post-crisis communication, including top officials of the Centers for Disease Control and Prevention during the anthrax crisis.

Shmuel C. Shapira, MD, MPH, is Deputy Director General of Hadassah University Hospital and Director of the School of Military Medicine at the Hebrew University–Hadassah School of Medicine in Jerusalem, Israel. An international authority on trauma, terror medicine, and emergency medicine, he has published extensively on these subjects and is co-editor of *Essentials of Terror Medicine* (Springer, 2009) and *Medical Response to Terror Threats* (IOS Press, 2010).

Samuel E. Shartar, BS, RN, CEN, is Senior Administrator of the Office of Critical Event Preparedness and Response (CEPAR) at Emory University. Before joining the CEPAR office, he was the Unit Director for the Emergency Department at Emory University Hospital.

Joshua Sinai, PhD, is an Associate Professor/Research, specializing in terrorism and counterterrorism studies, at the Virginia Tech Research Center in Arlington, Virginia. He previously worked on these issues in the US Department of Homeland Security and in the private sector, where, in his last position, he was detailed to work as a contractor at a government counterterrorism operations center.

Debra Wagner, CVA, NREMT, holds a BA in Criminal Justice and is the State Coordinator for the Oklahoma Medical Reserve Corps. She is a registered EMT and is one of about 1100 people worldwide to have earned professional certification in Volunteer Administration.

Panayotis A. Yannakogeorgos, PhD, is a Cyber Defense Analyst and Faculty Researcher at the Air Force Research Institute. He previously was Instructor and Senior Program Coordinator at the Division of Global Affairs, Rutgers University–Newark, where he was founding publisher and editor of the *Journal of Global Change and Governance*.

PROLOGUE

In July 2009 and September 2010, symposia on terror medicine and security were held under the auspices of the University of Medicine and Dentistry of New Jersey. The central aim was to give voice to experts in key fields involved with local preparedness, to assess the quality of preparedness and interaction among the fields, and to offer directions for improvement. Distinguished participants from the United States and Israel offered perspectives on a range of issues related to the field of terror medicine. This book has been developed from topics covered at the symposia.

The proceedings were immensely valuable to preparedness and response planning for mass casualty incidents, whether of natural, accidental, or terrorist cause. Support for the forums came from the highest levels of state government. New Jersey's Governor Chris Christie served as honorary chair of the 2010 symposium as did his predecessor, Governor Jon Corzine for the 2009 symposium. A keynote presentation was made by the Director of the New Jersey Office of Homeland Security and Preparedness: Charles McKenna in 2010 and Richard Cañas in 2009.

The symposia were also addressed by the Honorable Daniel Kurtzer, former United States Ambassador to Egypt and to Israel, and were supported by the Israeli Consulate in New York. Presentations were made at one or the other forum by prominent health and security leaders including Lt. Colonel Jerome Hatfield, Deputy Superintendent of Homeland Security, New Jersey State Police; Dr. Clifton Lacy, former New Jersey Commissioner of Health; Major General (res) Yitzak Gershon, former Head of the Israel Home Front Command; and Donald Jenkins, Col. USAF (ret), Mayo Clinic, former Medical Director of Trauma System for the US Central Command (including Afghanistan and Iraq).

One outcome was the familiarization of attendees with terror medicine. This emerging field encompasses aspects of emergency and disaster medicine as well as techniques for diagnosis, rescue, coordination, and security that are distinctive to a terrorist attack. The field is further discussed in Chapter 1.

Topics at the symposia ranged from treatment of injuries and emotional trauma to the role of the volunteer in a terrorist or disaster incident. At each symposium, experts from the United States and Israel addressed large audiences with a wide array of backgrounds. Attendees included physicians, nurses, dentists, paramedics, and others from the healthcare community, officials from law enforcement and security, and laypeople. Audience responses to the proceedings were uniformly enthusiastic.

The emphasis at the second forum was on local preparedness for terrorism and disaster, which is the focus of this book. Terrorism remains a global threat as evidenced by ongoing events in the Middle East, Europe, and beyond. The shootings in Fort Hood, Texas, in November 2009, were the largest terrorist attack on US soil since 9/11. Elsewhere in the United States, several planned attacks have failed or been thwarted. The continuing threat requires responsible awareness and preparation by both professionals and the general citizenry. Moreover, such preparation could be applicable to accidental or naturally caused disasters as well. Any individual could find himself at the scene of a terrorist or disaster event. With advanced preparation, any of us could be in a position to provide assistance to family members and other victims.

Chapter authors have been drawn from symposium presenters and others who represent a range of disciplines involved in local preparedness. The introductory chapters provide overviews of terror medicine, security, and communications, which are indispensable to successful preparedness. Subsequent chapters concentrate on a particular field and how responders from that field communicate and interact with others during and after an event. Thus, a chapter by a physician discusses not only the doctor's role but how that role is, or should be, coordinated with relevant others, such as emergency medical technicians and police.

Authors were asked to begin their chapters with a case study that demonstrates preparedness, or lack thereof, for a terrorist or disaster event. Their choices turned out to be fascinating and far ranging—from bioterrorism to earthquakes. Narratives of the case studies, often riveting, set the stage for further discussion from the perspective of the responder's field. Each author was asked to follow a common narrative sequence:

1. Recounting of the selected terrorist or disaster event
2. Description of preparedness and response efforts by the chapter's responder group
3. Manner of communication and interaction with other responder groups
4. Discussion of successes and failures of response efforts
5. Lessons learned including challenges, areas for improvement, and suggested next steps

The format was intended to help bind the book, with its many and varied contributors, into a coherent whole. The result, we believe, has been a successful compilation of different professional perspectives all linked by a singular purpose: to prepare for and respond to terror and disaster.

LAC
NDC

ACKNOWLEDGMENTS

This book is an outgrowth of symposia in 2009 and 2010 under the Program on Terror Medicine and Security, which is part of the Center for BioDefense at the University of Medicine and Dentistry of New Jersey. We are most grateful that Governor Chris Christie served as honorary chair of the 2010 Symposium, as did his predecessor Jon Corzine for the 2009 event, and for presentations by the Director of the New Jersey Office of Homeland Security, Charles McKenna in 2010 and Richard Cañas in 2009.

Daniel Kurtzer, former United States Ambassador to Egypt and to Israel, addressed both symposia, as did officials of the Israel Consulate in New York, Gil Lainer in 2010 and Benjamin Krasna in 2009. Several of the chapter authors in this volume were presenters at one or the other event, though we are indebted to all symposium participants for broadening our understanding of the emerging field of terror medicine and its relationship to security. Presenters at the 2010 Symposium included Dr. Bruria Adini, Erez Geller, Yitzhak Gershon, Jerome Hatfield, Rowena Madden, Tara Maffei, Brendan McCluskey, Dr. Mark Merlin, Dr. Yuri Millo, Dr. James Pruden, Megan Sullivan, and Sarri Singer. At the 2009 event: Dr. Isaac Ashkenazi, Michael Balboni, Henry Cortacans, Steven Crimando, John Grembowiec, David Gruber, Dr. Clifton Lacy, Dr. Jill Lipoti, Dr. Donald Jenkins, Dr. Tzipi Kahana, Rafi Ron, Estelle Rubinstein, Valerie Sellers, and Andrea Yonah.

Staff members of UMDNJ's Center for Continuing and Outreach Education were instrumental in promoting and organizing the symposia. For their unstinting support we thank in particular Patrick Dwyer, Theresa Setteducato, Jessica Young, and Attasha Nurse. From the UMDNJ Foundation, we acknowledge the dedicated efforts of George Heinrich and Elizabeth Ketterlinus.

Drs. Henry Falk and Isaac Ashekenazi, the authors of Chapter 10, wish to acknowledge colleagues whose work they referred to in their case studies and scenarios including Dr. Richard Hunt, Director of the Division of Injury Response at the CDC's National Center for Injury Prevention and Control, a strong sponsor of CDC guidance materials related to terrorism events and also of the Tale of Cities meetings; Drs. Leonard Marcus and Barry Dorn of the National Preparedness Leadership Institute at Harvard University; Madrid colleagues for their work toward understanding the response to the 2004 train bombings, and particularly Drs. Fernando Turegano-Fuentes and Ervigio Corral Torres for providing the figures in the chapter's depiction of those events.

Dr. Panayotis Yannakogeorgos, who authored Chapter 18, expresses his gratitude to Jennifer Lizzol for her contribution on the role of the Air Force during the Hurricane Katrina response. Others who read portions of the manuscript or otherwise provided thoughtful suggestions to enhance the project include Dr. David Baker, Team Leader of the Health Protection Agency of the United Kingdom; Howard Butt, State Coordinator for the Citizen Corps in New Jersey; Dr. Scott Compton, Assistant Dean for Educational Evaluation and Research, NJ Medical School; Rowena Madden, Executive Director of the NJ Governor's Office of Volunteerism; John Rollins, Specialist in Terrorism and National Security at the Congressional Research Service. Ruth Cole was an invaluable source of ideas and support at every stage of the project.

Finally, we are indebted to the talented staff at John Wiley Publishers, and especially to Karen Chambers, Life Sciences Editor, and Anna Ehler, Editorial Assistant, for their encouragement and guidance throughout the editorial process.

PART I

INTRODUCTION: THE KNOWN AND THE UNKNOWN

1

PREPAREDNESS, UNCERTAINTY, AND TERROR MEDICINE

LEONARD A. COLE

As mentioned in the Prologue, the chapters in this volume generally begin with a case study of a terrorist or disaster event. This narrative format introduces the main theme of the chapter as related to local preparedness.

In this chapter, the case study derives from attacks on July 7, 2005, when suicide terrorists bombed three London underground trains and one bus. The narrative here focuses on the bus bombing. Despite individual acts of heroism by responders and bystanders, organized response efforts were often wanting.

The lessons of 7/7, as the day is called, can be helpful in preparing for terrorist attacks in communities everywhere. Some of the experiences that day were unanticipated because they had seemed improbable. A range of uncertainties may apply as well to other events, whether they arise from deliberate, accidental, or natural causes. Among the lessons of 7/7 that apply to all such incidents is the importance of anticipating the unexpected.

THE BOMBING OF BUS NUMBER 30

Box-shaped stone homes grace the volcanic peaks of Santorini, an island 200 miles south of Greece. Amid the island's residences, blue-domed churches match the azure of the Aegean Sea a thousand feet below. Thought by some to be the ancient city of Atlantis, Santorini has long been a favorite of romantics.

Days after returning to London from a Santorini holiday, 28-year-old Sam and his fiancé Mandy still felt the island's glow. High on Sam's to-do list was a visit to Tiffany's to buy the promised ring for the woman he adored. Meanwhile, by Thursday morning, July 7, he had resumed his daily routine.[1]

Headed to the Central London office where he was a software specialist, Sam had settled into a window seat on the upper deck of the Number 30 (Figure 1.1). As the bus traveled

Local Planning for Terror and Disaster: From Bioterrorism to Earthquakes, First Edition.
Edited by Leonard A. Cole and Nancy D. Connell.
© 2012 John Wiley & Sons, Inc. Published 2012 by John Wiley & Sons, Inc.

FIGURE 1.1   Bus No. 30 on normal route to Central London. (Credit: Oxyman, Creative Commons Attribution 2.5 Generic License.) (For a color version of this figure, see the color plate section.)

east along Euston Road, it passed University College Hospital. An imposing structure of glass and steel, the hospital houses the largest critical care unit in Britain's National Health Service.

A block farther, beyond the red brick Quaker Friends House, traffic had become uncommonly dense. Diverted from its usual route on Euston, the Number 30 turned right and inched toward Tavistock Square. More commuters than usual were using buses and cars because train service had been suspended. The airwaves were carrying notice of a disruption to the city's underground rail system. It was approaching 9:45 A.M. and neither Sam nor his fellow passengers fully understood the cause of the congestion.

In fact, about an hour earlier, at 8:50 A.M., nearly simultaneous explosions had blown up trains in the London underground. As injured victims began to emerge from train stations, authorities suspected that a power surge had caused explosions at six stations. Only hours later, was it realized that three trains had been bombed.[2] The explosions had occurred deep in the tubes where the trains were en route between stations. Smoke and debris engulfed the passengers. Some lay dead, and others trapped. Many struggled to escape by foot along the tunnel tracks in either direction.

Four suicide terrorists, their backpacks laden with explosives, had planned simultaneous detonations on different train lines. One of them, Hassib Hussein, 18, found that service on the Northern Line train, his intended target, had been suspended. Unsure of his next step, he wandered for an hour. Upon reaching Euston Road, he boarded Bus Number 30 and moved toward the rear. He stopped directly beneath the seat occupied by Sam on the deck above.

Sam Ly and Mandy Ha had left Melbourne in 2003 for a 2-year working holiday in London. But home was still Australia. They had grown up there and planned to return soon for their wedding. Although they were part of Melbourne's Vietnamese community, their circle of friends extended far beyond. Sam, a graduate of Monash University, had worked in the school's Information Technology department. His reputation for patient assistance to computer-stumped students was legendary.

Sam's mother died of cancer when he was five, after which his father, Hi Ly, raised him alone. Father and son were still close. As Sam was riding to work, the picturesque card that he had sent from Santorini arrived in Hi Ly's mail, 10,000 miles away.[3]

Two long blocks south of Euston Road, Sam's bus reached the edge of Tavistock Square. To the right lay a small park whose centerpiece is a statue of Mahatma Ghandi sitting cross-legged in peaceful repose. To the left, set back from the curb, was the gated courtyard of BMA House, headquarters of the British Medical Association. BMA House stands on the site that once was home to Charles Dickens. It was there, in 1859, that he penned *A Tale of Two Cities*. "It was the best of times, it was the worst of times."

A mile and a half south, towering above Parliament, the hands of Big Ben glided to 9:47. The medical association's General Practitioners Committee was about to begin a meeting on BMA House's third floor. The half-dozen physicians in attendance heard a large bang. The room vibrated. Dr. Michelle Drage felt a heavy thud as a colleague jumped into her arms and drove them both to the floor.[4] According to Dr. Peter Holden, everything went "salmon pink."[5] (Drage and Holden recalled their experiences during hearings in 2011, part of the Coroner's Inquests into the events of 7/7. The hearings were presided over by Lady Justice Heather Hallett, a judge of the English Court of Appeal.[6])

In the bus outside, Hassib Hussein had detonated his backpack of acetone peroxide. The blast blew the roof off, crushed passengers, catapulted some to the road, and propelled severed arms and legs in every direction (Figure 1.2). All that could be seen from the third floor window of the BMA House was a fog of smoke. Doctors on lower floors went into the street but those at the meeting remained in the room. They deferred to Holden who had received training in prehospital emergency management. He worried about the possibility of additional explosions and advised that they not rush to the scene.

FIGURE 1.2 Bus No. 30 at Tavistock Square, London, after bombing on July 7, 2005. (Credit: Peter Macdiarmid/PA Wire URN:11139466. North American use only.) (For a color version of this figure, see the color plate section.)

Initial Response Efforts

During the next 15 minutes, the doctors in the street along with bystanders tried to remove the victims from the bus. Several passengers were dead and a few, barely alive, remained entrapped by twisted metal. But most of the casualties were carried off on makeshift stretchers, mainly tabletops (separated from the legs) obtained from the BMA House. They were placed on the courtyard ground or inside the building. When the doctors from the third floor went down to the street, patients were already spread across the courtyard. One of those physicians, James Dunn, described the situation as "fairly chaotic."[7] Dunn had reached the scene with Holden, who began to assume the role of incident medical commander.

Even with an abundance of doctors tending to the casualties, they could do little without equipment. No one even had a stethoscope and tablecloths were being used as bandages.[8] Michelle Drage recalled that for the most part, all they could do was offer reassurance to the victims.[9]

Dunn noticed that someone had obtained a bag of fluid, apparently saline solution, and was trying to insert a drip into an Asian young man lying in the courtyard. The man had a gaping right shoulder wound. Dunn went over to help and then stayed with him. As Dunn recalls, "He was fairly quiet all the way through the time I was looking after him and, at one stage, he shouted quite loudly, 'I just want to go to Australia.' After that, his level of consciousness did seem to decrease."[10]

Dunn worried that the man might be suffering from a brain injury or other internal damage. He conferred with Holden and both recognized that the patient was in urgent need of hospitalization. They agreed that he would be dispatched with the first ambulance to arrive. Meanwhile, it began to rain and the courtyard patients were carried on tabletops into the BMA building. Dunn remained inside with his patient whose name he eventually learned was Sam Ly. When an ambulance finally arrived, shortly before 11 A.M., Dunn had been waiting with Sam for an hour. At that point Euston Road had been cleared of nonessential traffic. The ride to University College Hospital, a few blocks away, took perhaps 3 minutes.[11]

By the time Sam was removed from the scene, 12 people had already died there. Seven died instantly from the force of the explosion, three in the next several minutes, and two later while lying in the courtyard. More than 100 other victims of the bus attack, some severely injured, continued to await removal.[12] Another hour and a half elapsed before enough ambulances had showed up to complete the evacuation. Thus, the bus site was finally cleared of victims $2\frac{1}{2}$ hours after the blast, and $3\frac{1}{2}$ hours after the first train was bombed.

Delays and Missteps

Might those whose conditions worsened during the delay, including the two who died in the courtyard, have benefited from rapid hospitalization? Colonel Peter Mahoney, a physician with the British Army, had led a team of experts to review information about some of the train and bus victims. Their sample included 18 who never made it to the hospital. (Sam Ly was not in the sample.) They concluded that 15 of them had suffered injuries that were "nonsurvivable." For the other three, the evidence was insufficient to reach a conclusion.[13]

Mahoney testified at the Coroner's Inquests in February 2011. Four months later, in her final report, Lady Justice Hallett, the presiding coroner, offered an opinion that reached beyond the evidence provided by Mahoney or anyone else. She determined "on the balance of probabilities that each of [the 52 mortally wounded victims from all the attacks] would have died whatever time the emergency services had reached and rescued them."[14]

Her presumption of "probabilities" appears subjective. One cannot know that rapid hospitalization would have made little difference for any of the victims. A hospital's range of diagnostic and therapeutic capabilities is manifold. It includes x-ray, MRI, and other imaging devices, drugs, oxygen, ventilators, specialists, operating facilities, all of which could only have been helpful. These features were unavailable to victims who, in some cases, waited hours before hospitalization. The "golden hour" is a tenet of emergency medicine. It refers not literally to 1 hour, but to the small window of time after injury when, for some patients, intervention can mean the difference between life and death.[15]

Actually, the ambulance that carried Sam had not been the first at the Tavistock location. About 10 minutes after the explosion, another one had completed a routine stop at the University College Hospital. Upon leaving the hospital and turning onto Euston Road, the driver and her fellow-paramedic noticed the commotion ahead. After taking some detours they reached the demolished bus. They parked close by and joined in moving victims to the courtyard. Then they unloaded their equipment, such as it was—a ventilation bag, collars for spinal injury, a few bags of saline fluid. Later, fearing that another explosive device might be on the bus, police blocked their return to the ambulance. The ambulance's paramedics then stayed to assist with patients on the ground.

The London Ambulance Service (LAS) is staffed by 5000 emergency medical and other support workers. Part of the National Health Service, its ambulances are distributed among 70 stations throughout the city. All requests from the public are directed to an Emergency Operations Center where a dispatcher checks the availability of ambulances near the persons in need.[16]

An ambulance might be sent to provide emergency aid and, if needed, transportation to a hospital. The designated hospital is commonly the closest one, or farther away if a trauma center or other special services seem necessary. If many casualties are involved, dispatchers must avoid overloading any single hospital with too many patients.

At 9:48 A.M., 1 minute after the explosion, the LAS received the first of several calls that a bus had blown up in Tavistock Square. Many people were screaming, said one caller. At least 10 were hurt, reported another. Yet as barrister Caoilfhionn Gallagher noted at subsequent hearings on the bombings, the warnings "were not acted upon."[17]

Terry Williamson, an operations manager with the ambulance service, testified that he and two crew members at the LAS headquarters on Waterloo Road were deployed to a supposed incident at the Liverpool Street tube station. Upon arrival they found the station empty and that no explosion had occurred there. Concluding that "there was no need for us to be there," they returned to headquarters and were tasked to Tavistock Square. Slowed by heavy traffic, the mile-plus drive from Waterloo took them 10 minutes. They had heard that an explosion had taken place at Tavistock but knew nothing more. No one told them that a bus was involved or that people were injured. This they learned when they arrived at Tavistock Square at 10:50, more than 1 hour after the blast there.[18]

Communications Failures

The crowd was thick outside the cordoned area and moving about was difficult. Worse, the radios worked sporadically or not at all. "We were unable to have communications with anyone," Williamson lamented—either at the scene or at central ambulance control. As a consequence, he underestimated the number of seriously wounded patients at Tavistock.

After 11 A.M., he spotted four newly arrived ambulances, but redirected three of them away. He thought the need was greater at a train station a few blocks south. Only at 11:20, a half hour after he had arrived at the Tavistock site, did Williamson learn that anyone had died there.[19]

Testimony from Dr. Tim Harris further underscored the devastating effects of inability to communicate. An emergency medicine physician, Harris worked at the Royal London Hospital. The Royal London is among six major hospitals within a half-mile of Tavistock Square. A leading trauma center, it is the base for the Helicopter Emergency Medical Service (HEMS) in which Harris also served. Several teams had already been sent to train stations when, shortly after 10 A.M., Harris learned about the Tavistock explosion. The HEMS's single helicopter and five rapid response cars are positioned to provide supplies and services by air or land. Their crews are trained to institute medical management when they arrive at any scene that contains a number of casualties.

Since Tavistock was nearby, Harris, with another physician and a paramedic, went there by car. They arrived with splints and other materials at 10:20, at which time Harris met Peter Holden. Harris was satisfied that Holden was overseeing medical care as well as possible under the circumstances. But he still lacked information about the availability of resources and the activities of other personnel. He tried to find the site commanders of other responder groups including police, fire, and ambulance services. He was unable to locate any of them face to face, by radio, or by phone.

Some responders had stopped using mobile phones out of fear that they could set off a secondary explosive. But whether for that reason or because phones had broken or the system was overloaded, every effort at distance communication had failed. Radios, mobile phones, the landline in the BMA House no longer functioned. Reliving his frustration, Harris recalled that "I was unable to access communications either to my parent hospital, to the coordinating desk of London HEMS, or to [command control] at the London Ambulance Service."[20]

Later at the coroner's hearing, toward the end of Terry Williamson's testimony, he was asked a summary question: "This entire episode . . . at Tavistock Square was bedeviled by failures of communications. Would that be fair?" To which he replied, "Yes," and emphatically, again, "Yes."[21]

In fact, communications and organizational mishaps had plagued response efforts not only at the bus site but throughout the city. Many individuals acted selflessly that day, but they were hampered by faulty information, confused directives, and poor coordination. The lapses were especially ironic at Tavistock Square, where a multitude of doctors was available within seconds of the blast, and hospitals were potentially reachable within minutes.

Before the day was over, Sam Ly had fallen into a coma and word of his condition eventually reached Australia. Hi Ly boarded a plane to be at his son's bedside while friends and family in Melbourne awaited updates on Sam's status. A friend from college, William Luu, posted a note on his Web site, *Will's Blog*: "I just don't know what to say or write right now. It's incredibly sad. I think the most I can do is just pray that he'll get through it okay."[22]

In the following days, dozens of friends offered comments and prayers on *Will's Blog*. The messages ranged from confident: "I remember he is a strong and bubbly character so I am sure he will wake up and come back to Melbourne"—to resigned: "I consider Sam one of the best friends a guy can have and I am saddened and shocked about what happened in London."

During the week, Sam was transferred to the National Hospital for Neurology and Neurosurgery, another of the hospitals within a half-mile of Tavistock Square. He died there on July 14, becoming the 13th fatality of the bus bombing and the 52nd of all the 7/7 victim fatalities. More than 700 others had been injured by the four bombings. Sam's fiancé and his father accompanied his body back to Australia. Before the burial, 100 members of Melbourne's Vietnamese community held a prayer vigil in celebration of his life.

What Went Wrong

On the eve of the London bombings, the United Kingdom would have seemed unusually well positioned to respond to a terror attack. The country's history of bombings by the Irish Republican Army (IRA) provided years of experience in dealing with terrorist assaults. Its medical, police, fire, and other responder personnel were highly skilled and sensitive to the threat.

The location of the bus explosion was especially fortuitous. Other than at a hospital entrance, it is hard to imagine a location better suited to help victims than at the front of a medical association headquarters. On 7/7, numerous doctors, though without equipment, were immediately available. Yet as later became evident, response efforts at that site were as wanting as at the underground train bombings. A sample of findings reveals a stunning assortment of deficiencies throughout the response network:

- Only half of the 201 London ambulances available in the area on July 7, 2005, were sent to the attack scenes.[23]
- Ambulance crews who were stationed near the attack sites were held back in case there were more assaults. Some of the crews were watching the events unfold on television.[24]
- Ambulance service headquarters was chaotic, with emergency phone calls going unanswered, key personnel unable to log on to computers, and emergency vehicle ignition keys lost.[25]
- At the ambulance control center, only one woman was tasked with logging all the emergency information as it was given to her on scraps of paper. She was in charge of updating the control room white board but could only reach half way up the board.[26]
- Eventually, there was so much information coming in to the control center that the staff there could not prioritize it effectively.[27]
- Firemen did not enter train tunnels because of their own safety rules even as injured passengers were walking through the tunnels toward exits.[28]
- Police officers discovered that their radios did not function underground.[29]
- Because of confusion and poor communications, fire and ambulance teams sometimes stood by while victims lay dying amid the train and bus wreckage.[30]
- Distribution of patients to some hospitals was hugely imbalanced. For example, some 200 patients were sent to the Royal London Hospital while Homerton, Newham, and Central Middlesex hospitals, which were all on standby, received none.[31]

A month after the bombings, officials of the London Ambulance Service authored an assessment titled, "Bombs Under London: The EMS Response Plan That Worked." The article displayed a self-congratulatory attitude that was common in the response community at that point. The authors' conclusion: "A tried and tested plan, well-trained crews and staff, and the availability of equipment on vehicles around London meant that London Ambulance Service was able to respond efficiently to these horrific events and maintain appropriate service levels to the rest of the city."[32]

As later became clear, the authors ignored or were unaware of numerous missteps. Some of these were acknowledged in a review ordered by the London Assembly, the city's oversight body of elected representatives. The assembly's July 7 Review Committee issued its report in June 2006. The "overarching, fundamental lesson," according to the report, was that procedures "tend[ed] to focus too much on incidents, rather than on individuals, and on processes rather than people."[33]

The document offered 54 recommendations. Some, such as calls to enhance communications among the responders and provide accurate information to the public, were obvious.[34] Other recommendations, while aiming to be helpful, seemed small bore. Among them, that first aid kits be placed on trains and in stations, that torches be available in case emergency lighting fails, that survivors be better informed about support services.[35]

A year later, the Review Committee issued a follow-up report indicating that several of its recommendations had been adopted, though for others, "more work is needed."[36] The Review Committee in 2006–2007 and the Coroner's Inquests in 2010–2011 provided a forum for hundreds of witnesses. Their testimonies led to a fuller understanding of the events of 7/7, and London has become better prepared as a result. But neither of the inquiries addressed a systemic weakness in London's preparedness, one that has been evident elsewhere as well. It derives from a rigidity of attitudes based on pre-event assumptions.

THE KNOWN, THE UNKNOWN, AND THE BLACK SWAN

Preparedness for terror and disaster logically takes into account past experience. But readiness for the known is, or should be, the easy part of the process. Preparing for the unknown is the greater challenge. In old Europe, people believed that all swans were white. They had never seen a black swan and assumed that none existed. In sixteenth century England, the term "black swan" was a common expression to signify that something was impossible. Only after the discovery of Australia and the sighting of actual black swans, was this age-old assumption abruptly ended. Nassim Nicholas Taleb employs the swan symbolism to describe effects of the highly improbable.[37]

Prior to September 11, 2001, the deliberate crashing of hijacked jetliners into buildings had never occurred in the United States or elsewhere. The seeming improbability of that scenario contributed to inadequate preparedness for its eventuality. Similarly, past experience with tsunamis prompted Japan to build 25-foot high coastal walls of protection. Then in March 2011, an earthquake unleashed a 30-foot wave that poured over the barriers and devastated the coastal lands. As with the 9/11 attacks, the Japanese assumption based on the previously known was upended by a *black swan*.

In the case of 7/7, response efforts were complicated by two deviations from the known. The first was the launching of nearly simultaneous attacks at different locations. The second was the nature of those locations. Before ending its armed campaign against the British in 2005, the IRA had launched hundreds of terrorist assaults. Almost all involved the detonation of a bomb at a single location. Each bombing was a discrete event. The biggest exception was on "Bloody Friday," July 21, 1972, when in an 80-minute period the IRA set off 22 bombs in Belfast. The explosions killed 9 people and injured 130.

As the years passed, Bloody Friday became a distant memory. Andrew Barr, a manager for the London Underground, recalled that during the 20 years prior to 2005, local responders had never drilled for "a multitude of attacks." Their exercises were limited to simulating an attack at one place.[38] The repeated single-bomb pattern primed British expectations and framed the country's approach to preparedness.

The second deviation involved the locations of the detonations. A few of the IRA assaults had been at underground train stations, though none on a train in transit. Thus, responders were more disposed to anticipate an explosion in a station than one deep in a tunnel between stations. When the improbable happened on 7/7, the authorities were confused. Hours passed before the conflicting reports were untangled. Contrary to the initial presumptions, survivors emerging from six stations were escaping from the aftermath of three explosions, and not six.

Other misconceptions persisted into the evening. Responders at the Tavistock Square bus bombing were quickly aware that at least seven victims had died immediately after the blast and that more died soon after. Yet an 8 P.M. BBC broadcast, 10 hours after the attack, reported a police affirmation that only two people had died there.[39]

The simultaneity and locations of the bombings were not responsible for every complication that day. But they contributed to key lapses, including communications failures. The net effect was prolonged uncertainty about where and how many explosions had occurred, how many casualties had resulted, and which locations needed priority attention.

After the attacks, perspectives about their likelihood sharply changed. The 7/7 scenario migrated from the realm of the improbable to that of the known. Since similar events were now deemed more likely, explicit efforts were implemented to prevent them. In fact, "preparing for the last event" is not unusual.

The phenomenon was patently demonstrated in recent experiences. Ever since Richard Reid tried in 2001 to ignite an explosive in his shoe during a flight, US air passengers have had to remove their shoes for inspection before boarding. Umar Abdul Mutallab's attempt in 2009 to detonate material hidden in his underwear subsequently led to more intrusive preboarding inspections of sensitive body areas.

But preparedness based only on past experience can give a false sense of security. Human ingenuity as well as natural forces have a way of circumventing the seemingly impregnable. A classic example was France's construction of the Maginot line along the German border in the early 1930s. Based on their experience with German troops in World War I, the French considered the fortification along its eastern front to be impassable. But in 1940, German ground forces simply flanked it by first moving through Belgium in the north and then pouring south into France.

Overconfidence about presumed preparedness is common as well with naturally occurring disasters. Severity of destruction is often unanticipated, as was the case with Hurricane Katrina in 2009, the Haiti earthquake in 2010, and the Japanese tsunami in 2011. In hindsight, the effects of these disasters could have been mitigated by better preparation. In each instance readiness had been based narrowly on assumptions from past experience.

Thus, sensible local preparedness cannot be rutted exclusively in the known. Of course, drills should largely be based on likely scenarios with responders acting in prescribed roles. But some exercises should also include imaginative eventualities. Even if a conjured scenario seems improbable, the effort has value. Creative thinking, whether about the known or the unknown, is a necessary part of preparedness.

In the months leading to 9/11, American planners barely recognized the threat of multiple hijacked aircraft. But during that period, in the mountains of Afghanistan, Osama bin Laden was leading a group of cohorts with a different idea. His Al Qaeda operatives were not only conjuring an "improbable" scenario but they were also about to make it happen. Similarly for the four jihadi terrorists who planned and implemented 7/7. A determination by the commission that investigated the September 11 attacks applies no less to 7/7. "The most important failure," the 9/11 Commission concluded, "was one of imagination."[40]

Of course, no one can anticipate every manner of assault. But exercises that include surprise scenarios can broaden the mindset of responders to better address the unexpected. The field of terror medicine offers a more expansive approach both to expectation and readiness.

TERROR MEDICINE

In May 2006, a year after the London bombings, Israeli authorities participated in a mock terror attack. The drill included more than a thousand personnel from the medical, police, fire, security, military, and other relevant communities. Participants knew in advance that a drill would take place at a certain time, but not about its nature or intended effects.

The exercise began with notification that a bomb had exploded in downtown Jerusalem resulting in numerous victims. Scores of mock patients began to arrive at the city's Hadassah hospital. They bore symptoms not only of the kinetic effects of the explosion, but also of some sort of chemical exposure. The hospital doctors and staff were challenged to identify the toxic agent and begin treatment. The victims reported their symptoms and held cards describing rashes that many were ostensibly developing. In less than 30 minutes, the doctors determined that the chemical was hydrogen fluoride and they began mock treatment accordingly. (I was an on-site observer of the drill.)

Especially interesting was not just the mixture of ingredients—a bomb coupled with a chemical agent—but the agent itself. Hydrogen fluoride, an industrial chemical that is corrosive to human tissue, is a precursor in the production of highly lethal nerve agents. It is recognized as an accidental-release hazard, though a literature search indicates no instance of this chemical having been released for hostile purposes. Thus, unlike sarin, mustard gas, and other familiar chemical agents, hydrogen fluoride would be a less expected choice. Using this unlikely chemical in the drill helped expand the responders' mindset beyond conventional expectations. Such thinking is but one aspect of terror medicine.

Terror medicine emerged as a distinctive medical field during the 5-year Palestinian uprising that began in 2000. In that period, Palestinians attempted some 20,000 terror attacks against Israelis.[41] More than 95% of them were thwarted, though still, about 1100 Israelis were killed and 6500 injured. The repeated attacks, including 150 suicide bombings, enabled Israelis to continually improve their techniques of rescue and treatment.

Among the concepts that arose from the Israeli experience were specific approaches to preparedness, incident management, treatment of injuries, and psychological consequences.[42] Israeli incident management, for example, became remarkably efficient. Although slower at the beginning of the uprising, toward the end, Israeli response time was remarkable. Following a terrorist bombing, 90% of the victims in need of hospitalization were commonly admitted within the first hour.[43] (Recall that after the London bus bombing 1 hour elapsed before even one of the victims reached a hospital.)

Features of terror medicine overlap with emergency and disaster medicine. But its singular characteristics range from treating exposures to a biological or chemical agent, like smallpox or sarin, to addressing the heightened emotional effects of a terrorist assault. Another important feature relates to the manner of dealing with multiple traumatic injuries.

Such injuries could include burns, bone fractures, ruptured organs, severed blood vessels, and other blast and crush effects. The chance that a physician would see all of these wounds in a single individual ordinarily is near zero. Yet after a close-quarter suicide bombing, dozens of victims may bear most or all of these injuries. Terror medicine includes determining which patients to treat first and in what manner. It also involves quick removal of a patient from an incident site to minimize his exposure to a possible secondary attack at that site.

Another aspect of terror medicine relates to security. Some Palestinians on their way to Israeli hospitals were found to have hidden explosives under their clothing. As a result, a policy of selective searches was implemented including of those in need of medical attention. Further, several Palestinian ambulances were discovered to have been carrying weapons and gunmen.[44] Now, all vehicles, including Israeli ambulances carrying people in urgent need of care, are stopped at the perimeter of a hospital's grounds. Only after the vehicle is inspected is it permitted to reach the entryway.

In the case of a natural disaster, the principal role of the security official is to maintain order. In a terrorist event, the role expands to include protection against a deliberate threat. A responder, whatever her field—medicine, administration, trained volunteer, police—is expected to cope with all such threats, whether emanating from terrorist, accidental, or natural causes.

ROADMAP FOR THIS BOOK

With appreciation for the breadth of a responder's roles, the introductory part of this book (Chapters 1–3) underscores the importance of communication and coordination during all such events. A sense of that importance continues as well through the book's three additional parts: the roles of health responders, of institutional managers, and of support and security personnel. Each part includes chapters by one or more experts in a specific field.

Every chapter's case study and follow-up assessment covers the quality of preparedness for that discipline. The chapters also make reference to coordination with responders from other fields. The aim is to discern the strengths and weaknesses of local preparedness from the vantage of key responder groups. As was clear in this chapter, good communication both within and between groups is central to the process. When medical responders to an event are well prepared but law enforcement responders are not—or vice versa—the outcome is jeopardized. Equally important, if responder groups are individually prepared, but not practiced at coordinating with each other, response efforts can be muddled.

The concluding chapter reviews the lessons provided throughout the book. It returns to a theme alluded to in this introductory section: preparation for the unknown as well the known. Approaching preparedness with an open mind and practiced interaction among responder groups could be essential to mitigating the effects of the next *black swan*.

NOTES

1. Details about Sam Ly's life were drawn from his friend's Web site and augmented by media reports. See "Will's Blog." July/August 2005. Available at http://will.id.au/blog/archive/2005/08,

accessed June 15, 2012; Wainwright M. Sam Ly. Guardian, August 20, 2005. Available at http://www.guardian.co.uk/uk/2005/aug/20/july7.uksecurity1, accessed June 15, 2012. Sam Ly: obituary. BBC News, updated January 20, 2011, http://www.bbc.co.uk/news/uk-12207397#story_continues_1, accessed June 15, 2012.

2. London blasts timeline. CBC News, July 22, 2005. Available at http://www.cbc.ca/news/background/london_bombing/timeline050707.html, accessed June 15, 2012.

3. Will's Blog; Wainwright; "Sam Ly: Obituary."

4. Hearing, Coroner's Inquests into the London Bombings of July 7, 2005. Judicial Communications Office. Available at http://7julyinquests.independent.gov.uk/hearing_transcripts/, accessed June 15, 2012. January 26, 2011, 131.

5. Hearing, January 28, 2011, 5.

6. Lady Justice Hallett heard testimony from 309 witnesses between November 2010 and March 2011.

7. Hearing, January 26, 2011, 5.

8. Hearing, January 25, 2011, 12.

9. Hearing, January 26, 2011, 136.

10. Hearing, January 26, 2011, 10.

11. Hearing, January 26, 2011, 64–70.

12. History, 7/7 London Bombings. Available at http://www.national911memorial.org/site/PageServer?pagename/new_history_impact_2005, accessed June 15, 2012.

13. Hearing, February 1, 2011, 28.

14. Hearing, May 6, 2011, 22.

15. Bohannon J. War as a laboratory for trauma research. Science 2011;331:1261–1262.

16. London Ambulance Service, Homepage, http://www.londonambulance.nhs.uk/about_us.aspx, accessed March 21, 2011.

17. Guardian.co.uk, January 31, 2011. http://www.guardian.co.uk/uk/2011/jan/31/77-inquiry-ambulance-delays/print, accessed March 15, 2011.

18. Hearing, January 28, 2011, 61–62.

19. Hearing, January 28, 2011, 65–69.

20. Hearing, January 28, 2011, 39–43.

21. Hearing, January 28, 2011, 71.

22. Will's Blog: http://will.id.au/blog/archive/2005/07/09/to-a-friend-hurt-in-london, accessed June 15, 2012.

23. Casciani D. How well did the 7/7 emergency services respond? BBC News, March 3, 2011. Available at http://www.bbc.co.uk/news/uk-12636731, accessed March 26, 2011.

24. Ibid.

25. Adley E. Messages were missed in ambulance hq chaos, 7/7 inquest hears. Guardian, February 28, 2011. Available at http://www.guardian.co.uk/uk/2011/feb/28/july-7-inquest-ambulance-chaos, accessed March 25, 2011.

26. "'Chaos' in 7/7 ambulance control hq." The Independent-UK, February 28, 2011. Available at http://www.independent.co.uk/news/uk/home-news/chaos-in-77-ambulance-control-hq-2228321.html, accessed March 25, 2011.

27. Ibid.

28. Doughty S. July 7 coroner launches scathing attack on emergency service "management jargon" that costs lives. Daily Mail, March 4, 2011. Available at http://www.dailymail.co.uk/news/article-1362655/July-7-coroner-launches-scathing-attack-emergency-service-management-jargon-costs-lives.html#ixzz1HYX9L5z1, accessed March 24, 2011.

29. Ibid.

30. Ibid.

31. July 7 ambulance "radio failure." BBC News, March 15, 2006. Available at http://news.bbc
.co.uk/2/hi/uk_news/england/london/4810368.stm, accessed March 26, 2011.

32. Hines S, Payne A, Edmondson J, Heightman AJ. Bombs under London: the EMS response plan
that worked. J Emerg Medical Services, 2005;30.

33. London Assembly, Report of the 7 July Review Committee. London: Greater London Authority, June 2006, 9. Available at http://legacy.london.gov.uk/assembly/reports/7july/report.pdf,
accessed May 18, 2011.

34. Ibid., 125–126, 135–136.

35. Ibid., 131–132, 139–140.

36. London Assembly, July 7 Review Committee, Follow-Up Report. London: Greater London Authority, August 2007, 8. Available at http://legacy.london.gov.uk/assembly/reports/7july/follow-
up-report.pdf, accessed May 18, 2011.

37. Taleb NN. The Black Swan: The Impact of the Highly Improbable, 2nd edn. New York: Random
House, 2010: xxi–xxiii.

38. Hearing, February 7, 2011, p. 10. Ironically, a private company exercise involving attacks at
several stations had been planned for the same day as the real bombings. See BBC Radio, July 7,
2005. Available at http://www.youtube.com/watch?v/sEbUQiYOGjU, accessed August 30, 2011.
Still, as indicated at the hearing, authorities had not in recent memory drilled for multi-point
attacks.

39. Metropolitan Police Deputy Assistant Commissioner Brian Paddick, cited in London bombing toll rises to 37. BBC, last updated July 7, 2005, 21: 50 P.M. Available at http://news
.bbc.co.uk/2/hi/uk_news/4661059.stm, accessed April 15, 2011.

40. The 9/11 Commission Report. National Commission on Terrorist Attacks Upon the United
States, Executive Summary, January 27, 2004, 7. Available at http://govinfo.library.unt.edu/
911/report/911Report_Exec.htm, accessed May 14, 2011.

41. Address by Israeli Prime Minister Ehud Olmert to the US Congress. Washington Post,
May 24, 2006. Available at http://www.washingtonpost.com/wp-dyn/content/article/2006/05/24/
AR2006052401420.html, accessed May 14, 2011.

42. Introduction to terror medicine, In: Shapira SC, Hammond JS, Cole LA, editors, *Essentials of
Terror Medicine*. New York: Springer; 2009, pp. 3–12.

43. The 90% figure has been drawn from debriefings following terror attacks, according to Dr.
Shmuel C. Shapira, Deputy Director General of Hadassah Hospital, per an e-mail communication,
May 15, 2011. Also, a timeline study of attacks against Israeli targets indicated that additional
ambulances were not usually needed after 40 minutes following an explosion, since by then
patients had already been sent off to hospitals. Aschkenasy-Steuer G, Shamir M, Rivkind A, et
al. Clinical Review: the Israeli experience: conventional terrorism and critical care. Critical Care
2005; 9: 490–499.

44. Use of ambulances by Palestinian terrorists. Available at http://www.standwithus.com/pdfs
/flyers/UNAmbulance.pdf, accessed June 15, 2012.

2

BIOTERRORISM AND THE COMMUNICATION OF UNCERTAINTY

Leonard A. Cole, Laura H. Kahn, and Peter M. Sandman

Biological weapons—pathogens used for hostile purposes—are different from any other category of weapons. A prime distinction is the fact that exposure to minute quantities of a biological agent may go unnoticed, yet ultimately be the cause of disease and death. The incubation period of a microbial agent can be days or weeks. Thus, unlike a bombing, knifing, or chemical dispersion, a bio-attack might not be recognized until long after the agent's release. Accordingly, bioterrorism poses unique challenges for preparedness and response. The challenges became patently clear during the nation's largest bioterrorism assault, in 2001.

THE 2001 ANTHRAX ATTACKS

In the wake of the jetliner attacks against the United States on 9/11, about a half-dozen letters containing anthrax spores were mailed to journalists and politicians. Four of the letters with spores and threat messages eventually were recovered. All were postmarked Trenton, New Jersey, which meant they had been processed at the postal distribution center in nearby Hamilton. Two letters were postmarked September 18, one addressed to Tom Brokaw at NBC-TV and another to the Editor of the *New York Post*. The other two letters were stamped October 9 and addressed to Senator Thomas Daschle and Senator Patrick Leahy.

As people became infected in September, October, and November, local responses revealed both strengths and gaps in preparedness for a biological attack. In the end at least 22 people had been infected, five of whom died. Meanwhile, scores of buildings had become contaminated with spores and more that 30,000 people who were deemed at risk required prophylactic antibiotics.[1] Millions more were fearful, many of them anxious about opening their own mail.

Local Planning for Terror and Disaster: From Bioterrorism to Earthquakes, First Edition.
Edited by Leonard A. Cole and Nancy D. Connell.
© 2012 John Wiley & Sons, Inc. Published 2012 by John Wiley & Sons, Inc.

The case study in this chapter focuses on the role-out of infections when their causes and appropriate responses were still uncertain. The first case of suspected anthrax was confirmed on October 4. The victim was a photojournalist at American Media, Inc. (AMI), a Florida tabloid publisher. He had contracted the inhalational form of the disease and died the day after his diagnosis was confirmed. Officials initially dismissed the possibility of a deliberate attack. "There is no evidence of terrorism," announced US Health and Human Services Secretary Tommy Thompson. "It appears that this is just an isolated case." Florida Governor Jeb Bush, noting that the incident occurred soon after 9/11, was emphatic: "People don't have any reason to be concerned. This is a cruel coincidence. That's all it is."[2]

Anthrax spores were later found on the photojournalist's computer and in the AMI mailroom, and a second AMI employee was diagnosed with anthrax. The peremptory dismissal of intentionality then seemed misplaced if not foolish. Still, no additional cases appeared during the next week, which gave hope that the outbreak had been limited to the two victims.

That all changed on October 12, when a black lesion on the skin of an NBC employee in New York City tested positive for anthrax. The NBC studios were immediately closed and the letter to Brokaw was later found in a pile of old mail. The next day, another case was belatedly confirmed. A producer at ABC's World News Tonight had taken her 7-month-old for a visit to the studio 2 weeks earlier. Soon after, a large lesion appeared on the baby's arm. Thinking the sore was caused by a spider bite, a doctor prescribed Benadryl, an antihistamine. The child's condition worsened, and on October 1, he was admitted to the New York University Medical Center. After 6 days of antibiotics his arm started to heal, but his red blood cell count dropped and his kidneys began to fail. "He was in deep trouble," his mother said. With continued antibiotics and blood transfusions, the child's condition began to improve, and a week later, he was released from the hospital.[3]

Meanwhile another anthrax front was opened on October 15, when an intern in Senator Daschle's office opened an envelope containing a small volume of powder. The accompanying letter read:

YOU CANNOT STOP US.
WE HAVE THIS ANTHRAX.
YOU DIE NOW.
ARE YOU AFRAID?
DEATH TO AMERICA.
DEATH TO ISRAEL.
ALLAH IS GREAT.

The frightened office staff was sequestered while the powder was analyzed. Hours later, laboratory tests indicated that the powder was indeed anthrax. Everyone in the vicinity of Daschle's office was given prophylactic antibiotics, and no one there subsequently became ill. But the story was different in several postal facilities.

The following narrative highlights the outbreak of illness among postal workers, the seeming improbability of spreading infection by mail, and the resulting confusion of communications among agencies and in messaging to the public.

THE MAIL AS A DISEASE CARRIER

On Friday, October 19, Joseph Curseen went to work for the last time.[4] Three days later he was dead. Curseen had helped operate machine number 17, a bar code sorter at the Brentwood postal distribution center in Washington, 2 miles northeast of the Capitol. He alternately fed letters into the machine and gathered them from its bins after they had been sorted electronically by bar code. The machine processed letters at the rate of 13,500 per hour. The process was brief though turbulent, as each letter was squeezed and buffeted through the sorter. No one at the time understood that the buffeting could force microscopic spores through an envelope's pores, which then, wafted by indoor air currents, spread throughout the building.

Thursday, October 18, 2001, was a pivotal day in the anthrax crisis. Although feeling out of sorts, Curseen had gone to work that day. Until then, four people were identified as having been infected with anthrax: the two AMI workers in Florida, the NBC employee, and the ABC producer's baby. On the basis of those cases, the risk seemed limited to people who had been near an opened letter. But additional findings that day raised new and larger worries. Lesions on a mail carrier and a postal mechanic in the Trenton area were identified as anthrax. Although postal employees had not previously been considered at risk, many were uneasy, especially at the Brentwood and Hamilton facilities.

To dispel anxiety, US Postmaster General John Potter decided to make an appearance at Brentwood. In the afternoon of October 18, he visited the center and mingled with employees. He assured them that they were not in danger. The Daschle letter, he said, "was extremely well sealed, and there is only a minute chance that anthrax spores escaped from it into the facility."[5] He was unaware of the infected postal workers in New Jersey or of those in Brentwood whose illness had yet to be identified.

Officials did not order the Brentwood facility closed until Sunday, October 21. But in New Jersey the decision to close Hamilton came earlier. Prompted by the news about the two Trenton-area postal workers, Dr. Eddy Bresnitz, the state's chief epidemiologist, visited the Hamilton distribution center. Three days earlier he had assured an assembly of workers there that they were not at risk. "The envelopes were sealed with tape," he told them, "and there's been no evidence of anthrax illness here." Now he had returned to announce that the building would be closed. Upon arrival he found that the machines were already shut down and no one was working. Postal Service officials were telling the press that the center would be closed pending additional testing.

That evening, Dr. George DiFerdinando, New Jersey's acting health commissioner, presided over a lengthy meeting with state health officials, postal authorities, and representatives of the Centers for Disease Control and Prevention (CDC). The group was trying to assess the situation and determine when the Hamilton center could be reopened. "We've hired a contractor who claims he can clean it . . . in somewhere between 24 and 48 hours," a postal representative said. "Well, how is he going to do this? What is his technique?" others asked. "We're not really sure," came the answer. Still, the postal people seemed to rely on the word of the contractor.[6]

The pressure to reopen Hamilton was driven in part by the huge loss of revenue and efficiency if mail had to be processed elsewhere. Further, the fact that the Brentwood facility was still open prompted some second guessing. "What do they know in Brentwood that we don't know, because they're not closing," one participant wondered. Jennita Reefhuis, one of the CDC officials present, empathized with the "postal guys," as she called them.

"They were so torn, because they saw our point about the need for safety, but they also had a business to run and knew that people were depending on their mail."

The discussion seemed endless until 1 A.M. when DiFerdinando called it to a halt. Whatever the postal service people decided to do or not to do, he announced, if they reopened that facility he would not be standing next to them. The room fell silent. "Well if you weren't standing there, we wouldn't be able to open," a postal representative said. DiFerdinando responded that he did not know what was going on in that building and could not vouch for its safety. Minutes later, everyone agreed that the facility would remain closed until more was known about the extent of contamination and how it would be cleaned up.

Eddy Bresnitz was at the meeting. Months later, in his office at the New Jersey State Department of Health and Senior Services, he considered what was subsequently learned about anthrax in the Hamilton postal center. Hanging on his wall was a large diagram of the facility's internal structure marked with hundreds of red dots. Each dot represented an area where anthrax spores were later found. No section of the 281,000 square foot structure was spared. Every corner from floor to ceiling was contaminated.

Bresnitz thought back to the meeting on the night of October 18 and the naïve optimism of the postal representatives. Fortunately, they eventually joined the consensus to keep the building shut. The diagram on the wall made clear that to have done otherwise could have been disastrous.[7]

On October 19, another Hamilton postal worker was found to have contracted skin anthrax and DiFerdinando again took the lead. Contrary to a CDC advisory, he declared that the state health department would now recommend that all 800 people who worked at the facility, and the 400 employees at the local post offices it serviced, begin preventive antibiotic treatment. No such early recommendation was made for the postal workers in Washington, where the Brentwood facility continued to operate.

In the days afterward, more postal workers at the Hamilton and Brentwood facilities were diagnosed with anthrax. Apparently, they had been infected prior to October 19, when New Jersey's broad antibiotic recommendation was issued. Some were critically ill, though the only ones who died were Curseen and Thomas Morris, both of them from Brentwood.

On November 2, the *Wall Street Journal* ran an article titled "Seven Days in October Spotlight Weakness of Bioterrorism Response." It reviewed the anthrax incidents between October 15 and 22 and noted the deaths of the two Brentwood postal workers.[8] Richard Custodio, a public health physician, summarized the article in *Medical Editor's Column*, an on-line publication of health management groups. He identified a single bright spot during that devastating week:

> At least one unsung medical hero emerges from this tragedy, Dr. DiFerdinando. He ignored CDC advice and gave the New Jersey postal workers Cipro. He took action with little infor-mation and remained answerable for his decision. His concern for safety, patient-centeredness and timeliness possibly saved lives.[9]

The two postal centers remained sealed for years. After several failed attempts, the Brentwood facility was decontaminated and reopened in December 2003. The Hamilton center was finally cleansed of spores and reopened in March 2005.

WHO ARE THE RISK COMMUNICATORS?

There are at least two leaders during public health crises such as the anthrax attacks: the elected official and the public health official. These two leaders need to communicate well with each other, with other officials, and with the public. When they do not work well together, problems may ensue.

Max Weber, the German sociologist of the late-nineteenth and early-twentieth century, believed that an inherent conflict exists between elected and appointed officials. Elected officials seek to get reelected in order to implement their ideologically based policies, while appointed officials strive to perpetuate and expand their bureaucracies. Weber described an inherent tension that exists between the two leaders.[10] This theory of conflict between elected and appointed officials is useful in studying leadership during crises such as bioterrorist attacks because how well officials lead and communicate can affect the outcome of a crisis.[11]

Political and bureaucratic leadership are different but equally important. In essence, there are two models of political leadership during a crisis: the Giuliani Model (named after former New York City Mayor Rudy Giuliani) and the Glendening Model (named after former Maryland Governor Parris Glendening). The Giuliani Model is exemplified by the relationship Mayor Giuliani had with New York City Health Commissioner Neal Cohen. During the anthrax crisis, Giuliani led by taking center stage communicating to the public through the press and by making key decisions with advice from Dr. Cohen. So in the Giuliani Model, the political leader takes charge and gets expert advice from his appointee.

Maryland Governor Parris Glendening led very differently. He preferred not to take center stage. Instead, he delegated communicating with the public and decision making to the expert, Dr. Georges Benjamin, the Maryland State Health Secretary. Benjamin made important policy decisions during the anthrax crisis, but he made sure to get political support from Governor Glendening. In the Glendening style of leadership, the political leader defers decision-making and public communication to the appointee. However, the political leader must give political support to the appointee by approving and supporting the appointee's policy decisions.

Both the Giuliani and Glendening models can work as long as the leaders understand their relationship, their roles, and responsibilities. Problems develop when these ground rules are not set. In New Jersey, Acting Governor Donald DiFrancesco and Acting State Health Commissioner Dr. George DiFerdinando had not had time to develop a working relationship when the anthrax attacks occurred. (DiFrancesco's predecessor, Christine Todd Whitman, had left state government months earlier to lead the US Environmental Protection Agency.)

DiFrancesco let Dr. DiFerdinando communicate to the public and make key decisions, but did not provide political support. Without political support, DiFerdinando ran into difficulties with the CDC when deciding to administer prophylactic antibiotics to the postal workers. The CDC disagreed with DiFerdinando's decision to treat the postal workers with antibiotics and refused him access to the national pharmaceutical stockpile.

Thus, DiFerdinando's advisory meant the workers would have to seek antibiotics on their own. On Friday evening, October 19, he went on television to urge the postal workers to see their physicians for the antibiotics. Unfortunately, many of these physicians were unavailable over the weekend and workers who managed to get prescriptions were unable to have them filled because pharmacies were running out of supplies.

Independently, Hamilton Mayor Glen Gilmore sought help from Christy Stephenson, the chief administrator of Hamilton's Robert Wood Johnson University Hospital. She ordered 18,000 pills from a pharmaceutical facility in southern New Jersey. The mayor sent a police car to pick them up and deliver them to the hospital, which served as an ad hoc public health facility for the postal workers. Hundreds of postal workers received antibiotics from this clinic.[12]

At the local level, Mayor Gilmore filled two roles: elected and public health official. There was no emergency medical infrastructure in place to provide care to the federal postal workers; it had to be improvised by the mayor and hospital administrator.

The anthrax attacks raised crisis risk communication to a new level. The attacks generated both a public health crisis and a high-profile criminal investigation. No one realized that spores could leak out of sealed envelopes and infect people.[13] No one was sure about who was at risk and what should be done to prevent infection in the different risk groups.

In the midst of uncertainty, DiFerdinando weighed the evidence and made the difficult decisions to close the Hamilton facility and recommend prophylactic antibiotics to the postal workers despite opposition by the CDC. The press vilified him, though it turned out that he had made the correct decisions after all.[14,15]

Problems arise when leaders hide or ignore a crisis, downplay its severity, give misleading or inaccurate information, or provide false reassurances. Journalists prefer to get information from scientists and medical professionals who can provide timely and accurate information. However, these professionals are not necessarily good public communicators since they tend to use technical jargon. Elected officials are usually competent communicators, but rarely have the expertise to accurately explain the crisis. They often give misinformation or false reassurances, as exemplified by Tommy Thompson and Jeb Bush at the outset of the anthrax crisis, noted at the beginning of this chapter.

Still, elected officials have the responsibility to communicate with their appointees and with the public. While the public expects honest, credible answers, even if it is bad news, it also expects that its elected officials are engaged and on top of the situation.

Ultimately, it is the elected officials who are in charge and responsible for the lives of their constituents. They decide whether to follow the Giuliani or Glendening model of political leadership. But whatever style of leadership they choose, it is up to them to ensure that effective and timely risk communication is conveyed to the public during a crisis.

COMMUNICATING RISK ABOUT BIOTERRORISM

Soon after the anthrax attacks, one of this chapter's authors (Sandman), produced an assessment titled "Anthrax, Bioterrorism, and Risk Communication: Guidelines for Action." The text was an expansion of a presentation he had made to the CDC. In it are several recommendations that are germane to events described in this chapter. Following are extended excerpts from that analysis.[16]

To make sense of my (Sandman's) recommendations for risk communication about bioterrorism, I must first outline two distinctions fundamental to my approach: hazard versus outrage, and public relations versus stakeholder relations. Bioterrorism is atypical with respect to both of these distinctions, in ways that make risk communication about bioterrorism different from most risk communication.

The factors that determine the public's response to a risk, it turns out, are genuine characteristics of that risk. But they have very little to do with its associated mortality, morbidity, or ecosystem damage. These are what the experts respond to. The public responds far more to factors like the risk's voluntariness, control, dread, familiarity, and memorability; and to the extent to which the institutions associated with that risk seem trustworthy and responsive. I usually call these "the outrage factors." I distinguish two aspects of a risk:

its hazard, which determines the experts' response; and its outrage, which determines the public's response. When outrage is high, people will be concerned, independent of hazard. And when outrage is low, they will not, again independent of hazard.

It follows that when risk communicators are trying to warn people, one of the main things they do is try to increase the outrage. And when risk communicators are trying to reassure people, reducing the outrage is a key task.

How Is Bioterrorism Different?

In the anthrax attacks of late 2001, these two distinctions (hazard vs. outrage and PR vs. stakeholder relations) played out in unusual ways—not unique, but unusual enough to require adjustment to the conventional strategies of risk communication. I suspect the same will be true of future bioterrorism risks.

Any use of biological agents as weapons of mass destruction—in fact, anything even close to that—will instantly turn the public into stakeholders. Even comparatively unconcerned people will be concerned enough to pay attention, watching for signals that they ought to hike their level of concern. Most people will be vigilant. Many will be hyper-vigilant, spending their days and even their nights with CNN. And the people who are not paying attention may not be apathetic at all, but something much more complicated: too terrified to watch, or shocked into denial.

What happens to the hazard-versus-outrage distinction during a bioterrorist attack? Usually, the range of possible reactions to a risk runs from apathy at one extreme to outrage at the other extreme. "Outrage" includes fear, anger, concern, etc. These emotions are usually experienced without distortion; outraged people know they are outraged, and know what they are outraged about. But sometimes—and a bioterrorist attack is surely one of those times—the level of concern moves beyond normal outrage. One possible version of this "extreme beyond the extreme" is panic. A more common version (because panic is relatively rare) is denial. Risk communicators do not usually need to think too much about how to address panic and denial; apathy and outrage are our daily adversaries.

Recommendations to Risk Communicators[17]

Do Not Overreassure   People are ambivalent about bioterrorism risks. Bioterrorism is high outrage: It is (among other outrage characteristics) catastrophic, unknowable, dreaded, in someone else's control, morally relevant, and memorable. Yet people recognize that their personal risk, statistically, is quite low so far. Hence, the ambivalence. So reassuring them—riding the confident seat on the seesaw—backfires, it forces the public onto the worried seat.

The paradoxical intervention is the one that works. Tell people how scary the situation is, even though the actual numbers are small. And watch them get calmer.

Acknowledge Uncertainty   Even if unacknowledged, uncertainty [usually] provokes very little criticism of people in authority during a crisis; we are too dependent on [their] protection at that point to dare to question [their] competence. But when the crisis is over, the bill comes due: Inevitable stories about what [was] mishandled or nearly mishandled. (The most hostile media criticism of CDC's handling of the 2001 anthrax attacks was probably the coverage of its uncertainty about what exposed-but-not-sick patients should do when their 60-day course of antibiotics was over. Significantly, the story emerged weeks after the last anthrax case was found.) Track any risk crisis, from Three Mile Island in 1979 to anthrax in 2001, and you will find the critical assessments of crisis management begin a few days after the crisis itself abates, and build from there.

We want our leaders to be confident but not overconfident in the face of uncertainty. We want them to tell us the unpleasant truth about the situation's uncertainties and the need to act without all the facts—but we want them to bear this truth and help us bear it too. What we most value in our leaders, in fact, is the ability to function in ambiguous realities without collapsing into timidity and without arrogantly wishing away the ambiguities.

Share Dilemmas   Dilemma sharing is hard because it goes against everybody's grain. Scientific sources usually would rather reach their best judgment, however tentative, and then claim confidence. And the audience usually would rather be told what to do by a confident scientific source. We all tend to get overly dependent in a medical emergency, whether the crisis is personal or a matter of public health. Unfortunately, that does not keep us from wreaking vengeance if an expert gives us overconfident advice that turns out bad. For the expert, therefore, the choice is clear: Irritate your audience now by acknowledging uncertainty and sharing the dilemma; or claim omniscience now and risk paying a far higher cost later in outrage and lost credibility.

[An example during the anthrax event] was what to do with the cohort of individuals who were just finishing their 60-day course of antibiotics after possibly being exposed to anthrax. CDC had based the original 60-day regimen on the only research it had—decades-old research on natural anthrax and healthy patients. But there was animal research suggesting that anthrax spores might conceivably survive in the lungs for longer than 60 days. And CDC scientists had learned to expect the unexpected.

So CDC offered patients a choice among three options: (a) Stop at 60 days, and watch for the remote possibility of illness; if it happens, get to a doctor fast. (b) Take another 40 days of antibiotics, trading the additional risk of antibiotic side effects for the additional protection against any anthrax spores that might still be lurking. (c) Take the anthrax vaccine—which has possible side effects of its own, and has never been tested on people already exposed to the disease. CDC made no recommendation. Any of the three options, it said, had low, uncertain, non-zero risk; given the sorry state of the science, any of the three might be best.

The reaction from the media, from politicians, and (at least as quoted in the media) from the patients themselves was uniformly negative. CDC's clear statement about the unclear nature of the science was described as "muddled" and "confused"; CDC was repeatedly characterized as having "admitted" that it was not sure what patients should do, as if being sure, or sounding sure, were its obvious responsibility. In a congratulations-and-condolences email to CDC officials, I wrote: "There is this consolation: The criticism that comes to an agency that refuses to give advice when it lacks a scientific basis is nothing compared to the criticism that comes to an agency that guesses wrong."

Give People a Choice of Actions to Match Their Level of Concern   The best risk communication gives people a choice of things to do—the act of choosing is itself an assertion of control that binds anxiety, prevents panic, and reduces denial. Ideally, your menu of protective responses ranges around a recommended middle. "X is the minimum precaution; at least do X. Y is more protective, and we think wiser; we recommend Y. Z is more protective still, and we think a little excessive—but if you're especially vulnerable or especially concerned, by all means go that extra mile and do Z."

The X-Y-Z choice tells people how concerned you think they ought to be, the level of concern represented by protective response Y. But it also gives people permission to be more or less concerned than you think they ought to be—and for whatever level of concern they are experiencing, it prescribes a set of precautions.

It is not rare—in fact, it is the norm—to find yourself in a situation where you know enough to give advice but not enough to insist that any behavior other than your recommended behavior is flat-out wrong. Yet some behaviors are flat-out wrong. So you prescribe a range: your recommendation surrounded by less and more protective strategies that you consider not unreasonable. You must do this, of course, in a way that does not send a double-message. "It's nothing to worry about but take antibiotics anyway" is a double-message. "It's probably nothing to worry about, but if you want to take antibiotics just to be on the safe side, that's okay too" is not.

CDC Director Jeffrey Koplan provided a superb example in a November 4, 2001, interview with Wolf Blitzer on CNN. Koplan was explaining that CDC now recommended that many people who had been started on antibiotics should stop, because it had been determined that their exposure was negligible after all. But he gave them permission to ignore the recommendation and keep taking the meds for the full 60 days: "Some of them may decide that, for the sense of their own security, they want to. That's a decision they may make." Paradoxically, it is easier to decide to come off of Cipro if CDC says that is what it recommends, but if you are really worried, go ahead and stay on it.

CONCLUSION AND SALIENT OBSERVATIONS

- The first confirmation of an anthrax case was on October 4, 2001, but between September 21 and October 1, nine people had shown symptoms of the disease (seven cutaneous and two inhalation). Their illness had been misidentified because of faulty diagnoses or erroneous laboratory tests. Accurate identification came weeks or months later, after retrospective assessments and retesting.

- After the Hamilton, New Jersey postal center was closed because of possible anthrax contamination, employees there wondered why the Brentwood facility in Washington remained open. Conversely, employees at Brentwood wondered why their facility was still operating when Hamilton was closed.

- Contrary to the initial plans of postal authorities, the Hamilton postal center remained closed indefinitely. And contrary to advice from the CDC, New Jersey health officials recommended that every postal employee in the Hamilton–Trenton area take preventive antibiotics.

- Prior to the 2001 anthrax events, many bio-attack scenarios had been conjured, which ranged from the release of pathogens from an airplane to spraying them from an aerosol canister. Plans never included the possibility of spreading microbes through mailed letters. The improbable had occurred: a *black swan*.[18]

- Coordination and communication lapses were attributable in part to incomplete knowledge about the danger posed by an attack launched via the US mail.

- Authorities initially minimized or dismissed the possibility that the first confirmed anthrax case (on October 4) was deliberately caused. When shown to be wrong, their premature assurances could only have weakened public confidence in further official pronouncements.

- Despite official assurances that the mail was safe, news reports and surveys indicated that many people were afraid to open their mail.
- Distinctive features about bioterrorism require different approaches to communication and messaging than do other forms of terrorism.

NOTES

1. Jernigan DB, Raghunathan PL, Bell BP, et al. Investigation of bioterrorism-related anthrax, United States, 2001: Epidemiologic findings. Emerg Infect Dis 2002; 8(10):1019–1028.
2. Bhatt S. "No evidence of terrorism" in isolated case, US Health Secretary says. The Palm Beach Post, October 5, 2001.
3. Lipton E. Taking baby to the office, then living a nightmare. The New York Times, October 17, 2001; B-6.
4. Much of this narrative is drawn from Cole LA. *The Anthrax Letters: A Bioterrorism Expert Investigates the Attacks That Shocked America. Revised.* New York: Skyhorse Horse Publishing; 2009; earlier version: Cole LA. *The Anthrax Letters: A Medical Detective Story.* Washington, DC: Joseph Henry Press/National Academies Press; 2003.
5. Postmaster General John E. Potter, Press Conference at Brentwood Mail Processing and Distribution Center, Washington, DC, October 18, 2001.
6. Reports of the meeting were recounted in interviews with attendees including NJ. Acting Health Commissioner George DiFerdinando, NJ State Epidemiologist Eddy Bresnitz, and CDC representatives Jennita Reefuis and Christina Tan. The interviews were conducted by Leonard Cole in 2001–2002.
7. Anthrax lethality depends on the number of spores inhaled and the susceptibility of the exposed individual. For some people, fewer than 100 spores presumably could be deadly: "The most widely accepted estimate of inhaled spores required to produce a lethal dose in 50 percent of the population is 8,000. However, researchers at the University of Texas Medical Center using 'probit' models, estimate that only 98 inhaled spores may cause lethal infection in 10 percent of the population." See more deadly anthrax spores may remain after decontamination. Science Daily. October 25, 2005. Available at http://www.sciencedaily.com/releases/2005/10/051026082623.htm, accessed June 16, 2012.
8. Chen K, Hitt G, McGinley L, Peterson A. Seven days in October spotlight weakness of US response to threat of bioterrorism. The Wall Street Journal. November 2, 2001. Available at http://www.ph.ucla.edu/epi/bioter/pastweakness.html, accessed June 16, 2012.
9. Cited in Cole, the Anthrax Letters: A Bioterrorism Expert; 81.
10. Weber M. Bureaucracy. In: Gerth HH, Mills CR, editors. *Max Weber: Essays in Sociology.* New York: Oxford University Press; 1946. pp. 228–244.
11. Kahn LH. *Who's in Charge? Leadership During Epidemics, Bioterror Attacks, and Other Public Health Crises.* Santa Barbara, CA: Praeger Security International; 2009.
12. Winerip M. Our towns; hail the mayor (whose name isn't Giuliani). The New York Times. November 14, 2001;14. Available at http://www.nytimes.com/2001/11/14/nyregion/our-towns-hail-the-mayor-whose-name-isn-t-giuliani.html, accessed June 16, 2012.
13. Ironically, just prior to the anthrax attacks, a Canadian study was performed in which envelopes containing bacterial spores were opened in a confined area. The test bacteria were *Bacillus subtilis*, a relatively harmless simulant of *Bacillus anthracis*. After the envelopes were torn open, the spores quickly spread and anyone in the area would have been heavily exposed. Kournikakis B, Armour SJ, Boulet CA, Spence M, Parsons B. Risk assessment of anthrax threat letters. Defence Research Establishment, Suffield, Canada. Technical Report, DRES TR-2001-048, September 2001.

14. Winerip.

15. Kleinfield NR. A nation challenged: the sites; anthrax closes a third New Jersey post office. The New York Times. October 28, 2001. Available at http://www.nytimes.com/2001/10/28/nyregion/a-nation-challenged-the-sites-anthrax-closes-a-3rd-new-jersey-post-office.html, accessed June 16, 2012.

16. For the complete article see Sandman PM. Anthrax, bioterrorism, and risk communication: guidelines for action. The Peter Sandman Risk Communication Website, December 29, 2001. Available at http://www.psandman.com/col/part1.htm, accessed June 16, 2012.

17. Ibid. Sandman's article included 26 recommendations to enhance risk communication. The four cited here are especially relevant to the narrative in this chapter.

18. For more on the black swan metaphor, see Chapters 1 and 19.

3

RESPONDING TO DISASTER AND TERRORISM: THE CENTRAL ROLE OF COMMUNICATION

STEVEN M. BECKER

Many elements are important in the response to a disaster or terrorist incident. The ability of people in affected areas to quickly undertake self-protective actions can be vital. So, too, can actions by responders to rapidly screen a population to identify those at risk for health effects. But regardless of which elements are most salient in a particular crisis situation, they all depend on effective emergency communication and information. Indeed, effectively communicating with people and understanding critical information needs can mean the difference between success and failure in the response to any disaster or terrorist event.

LEARNING FROM PAST INCIDENTS

The Chernobyl disaster, which took place in the former Soviet Union, provides a powerful illustration of communication's importance (Figure 3.1). The Chernobyl power station, located 100 km from Ukraine's capital Kiev, was home to four large nuclear reactors. On April 26, 1986, a power surge in the number four reactor resulted in a massive steam explosion, the rupture of the reactor vessel, and a fire that burned for 10 days. The reactor did not have a concrete and steel containment structure, and massive quantities of radioactive materials were released over parts of Europe, particularly Ukraine, Belarus, and Russia. Ultimately, some 345,000 people were evacuated and resettled in the aftermath of the accident, an experience that proved deeply traumatic.[1]

Cows grazed in pastures where radioactive iodine had settled, resulting in highly contaminated milk. Although authorities took some measures to reduce exposure and health risks, these were inadequate because many farmers and members of the public were not given

Local Planning for Terror and Disaster: From Bioterrorism to Earthquakes, First Edition.
Edited by Leonard A. Cole and Nancy D. Connell.
© 2012 John Wiley & Sons, Inc. Published 2012 by John Wiley & Sons, Inc.

FIGURE 3.1 Chernobyl Nuclear Power Plant, post-1986. (Photo credit: SM Becker.) (For a color version of this figure, see the color plate section.)

"timely information about the accident and necessary responses."[2] As a result, children drank contaminated milk, which exposed them to high levels of radioactive iodine.

Iodine concentrates in the thyroid, a gland in the front of the neck that plays an important part in metabolism. The thyroid cannot distinguish between regular iodine and radioactive iodine; thus, when children drank the contaminated milk, their thyroid glands absorbed the radioactive element. Furthermore, because children in parts of the former Soviet Union already had insufficient amounts of iodine in their diets, their thyroids took in even greater quantities of the radioactive isotope.

The result has been a sharp increase in childhood thyroid cancer since the disaster: "Amongst those under age 14 years in 1986, 5,127 cases (under age 18 years in 1986, 6,848 cases) of thyroid cancer were reported between 1991 and 2005."[3] Drinking milk "from cows that ate contaminated grass . . . was one of the main reasons for the high doses to the thyroid of children, and why so many children subsequently developed thyroid cancer."[4]

If proper information had been communicated in a timely fashion, the results could have been different. "Since radioactive iodine is short lived, if people had stopped giving locally supplied contaminated milk to children for a few months following the accident, it is likely that most of the increase in radiation induced thyroid cancer would not have resulted."[5]

Chernobyl dramatically demonstrates how inadequate emergency communication can result in long-term health risks. Poor communication and information practices can also translate into an immediate increase in mortality. The Bhopal chemical disaster is a compelling case in point. On the night of December 2, 1984, some 40 tons of methyl isocyanate (MIC) gas escaped from a Union Carbide plant in Bhopal, India. MIC is a very reactive, highly toxic chemical used in pesticide production. The spreading gas reached hundreds of thousands of people. Many suffered ocular and pulmonary damage, experiencing such problems as burning in the eyes, breathlessness, choking and wheezing, and pulmonary edema. Thousands of people died.[6–8]

The disaster's impacts were worsened by many factors, including the failure of critical plant safety systems, high population density near the plant, atmospheric conditions that kept the chemical cloud close to the ground, and the fact that most people were asleep when the release occurred. But communication and information failures also contributed substantially to the morbidity and mortality. Area residents, lacking information about the chemicals utilized at the plant, had no idea what to do. Already suffering from exposure to the chemical, many simply tried to get away. "In the darkness, victims suffering from the effects of MIC were unable to determine the precise source of the toxic plume; unfortunately, some moved toward the source of MIC. . . ."[9] Meanwhile, the emergency services, which also lacked information about the chemical and appropriate protective measures, tried to help by urging people to flee.[10] But this was exactly the wrong action to take.

> [N]ot knowing what to do, many people chose to flee and thus were exposed to the methyl isocyanate gas. If these people had been informed in a timely manner, they may have known that the best emergency protection against the chemical was to lie on the floor in an enclosed space with wet cloths on one's face. Many lives would probably have been saved if people had been aware of such a simple safety measure.[11]

Chernobyl and Bhopal serve as stark reminders of the importance of emergency communication. In both cases, citizens were not provided with information needed to protect themselves and their loved ones. In one case, the result was long-term health effects (cancer); in the other, an immediate increase in mortality.

Chernobyl and Bhopal were major crises, and not all incidents threaten the health of so many people. But in all disasters and terrorism incidents, communication is central. "Timely, accurate information can help people at risk take appropriate protective measures" and this, in turn, can help "prevent illness and injury."[12] Indeed, in many situations, the single most important public health intervention is rapid and effective communication of protective action information. Thus, communication issues need to be at the heart of preparedness and response efforts. In practical terms, this means not only incorporating them into plans but also holding periodic exercises that rehearse and test what is in the plans.

THE MANY FACETS OF EMERGENCY COMMUNICTION: JAPAN 2011

The examples discussed above underline the importance of delivering timely self-protection information to the public. But other communication needs can also be crucial in a crisis situation. To shed light on the broad range of communication and information issues that can arise in a disaster or terrorism incident, it is useful to consider the Fukushima Dai-ichi nuclear crisis that occurred in Japan in 2011.

On March 11 of that year, at 2:46 P.M. local time, a huge earthquake occurred off the east coast of Japan. The magnitude 9.0 quake was the biggest ever to hit modern Japan, and one of the five largest in the world, since scientific record-keeping began. In addition to the direct damage it caused, the earthquake produced a massive tsunami that swept deep

FIGURE 3.2 Japanese police guarding an entrance to the 20 km "no go" zone around the Fukushima Dai-ichi Nuclear Power Plant, 2011. (Photo credit: SM Becker.) (For a color version of this figure, see the color plate section.)

inland. The disaster killed more than 15,800 people, left thousands of others missing, and caused widespread destruction. In addition, the calamity seriously damaged the Fukushima Dai-ichi facility, one of the world's largest nuclear power stations. Large quantities of radioactive materials were released, and more than 100,000 people in the vicinity evacuated (Figure 3.2).

Japanese society demonstrated remarkable courage and resilience. "Large numbers of people . . . even in the face of a triple disaster, horrendous losses, and difficult long-term challenges . . . exhibited dignity, resolve and compassion."[13]

> [I]n evacuation shelters . . . an atmosphere of civility prevailed despite difficult conditions and deep uncertainty about the future. In other locations, people organized campaigns to help farmers whose livelihoods had been damaged, or collected food and other necessities for evacuees. . . . Throughout Japan, individuals volunteered by the tens of thousands to do everything from cleaning up disaster damage to providing healthcare.[14]

From the outset of the nuclear crisis, communication issues moved center stage. Whether near or far from the plant, the demand for information soared. The most notable issues were discussed in the following sections.

The Public Wanted Information

As the crisis at Fukushima Dai-ichi unfolded, Japanese authorities found themselves in uncharted territory. Their immediate focus was on managing the developing emergency to prevent an even bigger calamity. Authorities moved quickly to reduce potential public exposure to radiation and radioactive contamination. One major action was to evacuate people from communities in proximity to the crippled facility. Another step involved stopping the distribution of raw milk, spinach, and *kakina* (a leafy vegetable) to minimize consumption of radioactive iodine.[15] Such measures helped to reduce potential public health impacts.

Not surprisingly, from the start of the crisis, there was a great demand for information. Yet, for several reasons, authorities had difficulty responding effectively in the early stages of the disaster. First, there was the challenge of having to deal with the nuclear crisis while still reeling from the earthquake and tsunami. As Japanese officials commented in a report to the International Atomic Energy Agency (IAEA): "The situation has become extremely trying for Japan, insofar as it has had to execute countermeasures for the nuclear accident while also dealing with the broader disaster caused by the earthquake and tsunamis." Never having faced a situation of this type, utility officials and government authorities alike had their hands full. Furthermore, the natural disasters damaged the communications infrastructure, hindering attempts to reach people: "Communication to residents in the surrounding area was difficult because communication tools were damaged by the large-scale earthquake."[16]

Second, as Japanese agencies noted in their own reviews of response operations, communication and information efforts were hampered by major management and coordination problems. In particular, the roles and responsibilities of the various groups managing the disaster were not clearly defined—between local and national emergency response offices, between the government and the utility, or among governmental agencies. In addition, "communication was not sufficient between the government and the TEPCO [Tokyo Electric Power Company] as the accident initially began to unfold." The net effect was that at times no information was forthcoming, and at other times seemingly inconsistent information was issued.[17]

For local governments and the public in affected areas, this meant that important information "was not always provided in a timely manner." In addition, health implications and radiological protection guidelines "were not sufficiently explained."[18] Mixed messages including conflicting assessments of how long people would need to stay away added further to confusion and anxiety. Meanwhile, there was criticism from the media, local officials, and the public over something known as SPEEDI—the System for Prediction of Environmental Emergency Dose Information. SPEEDI helps authorities model the dispersion of radioactive materials. In the early stages of the disaster, data generated by SPEEDI were not disclosed to local governments or the public. This appears to have been due to a combination of factors, ranging from differing assessments of the data's reliability to breakdowns in interagency communication. The net result was that potentially valuable dispersion information, particularly about the direction of the plume, was not available to inform the evacuation process. According to both Japanese and international reports, this resulted in some people evacuating from less contaminated areas to areas that were in the path of radioactive releases.[19,20]

In several reports reviewing the management of the accident, Japanese officials were candid in recognizing the seriousness of the communication problems: "Especially immediately after this accident, actions were not sufficiently taken to provide local residents with information or easily understood explanations about radiation, radioactive materials, or information on future outlooks on risk factors."[14] In addition, "although the results generated by SPEEDI are now being disclosed, disclosure should have been conducted from the initial stage."[21] This also could have led to safer evacuation directions.

The importance of communication lapses was further underscored by an independent investigative committee created by the Japanese parliament. In the executive summary of its December 2011 preliminary report, the committee commented:

> The following tendency was observed: transmission and public announcement of information on urgent matter was delayed, press releases were withheld, and explanations were kept ambiguous. Whatever the reasons behind, such tendency was hardly appropriate, in view of communication in an emergency.[22]

But government was by no means the only actor found wanting in terms of communication. Criticism was also leveled by the public, the media, and the local government at TEPCO, the utility that operated the Fukushima Dai-ichi plant. The public had great admiration for the courage and dedication of the TEPCO workers who were battling under very difficult conditions to bring the damaged nuclear facility under control. But from early on, the company and its spokespersons were accused of not being forthcoming or of only providing vague or inadequate information. Perhaps hoping to maintain calm, TEPCO's descriptions of events were sometimes measured or euphemistic. For example, on March 12, a huge hydrogen explosion at the Unit 1 reactor blew off the facility's outer walls, but a company press release indicated only that "there was a big sound around the Unit 1 and white smoke."[23] The gulf between TEPCO's descriptions and those of the media and the public only served to increase people's distrust.

What was seen by some as excessive secrecy was also criticized. When a parliamentary committee asked the utility for copies of two operating manuals, it initially received only one which was largely blacked out. Later, when the second manual was provided, it, too, was almost entirely redacted.[24] The utility maintained that the materials were internal documents and not for general publication. TEPCO claimed the redactions were necessary to protect both intellectual property rights and nuclear materials.

The parliamentary committee, however, rejected the company's claims: "It is unacceptable for TEPCO to refuse to disclose these materials in the wake of this kind of disaster," committee chair Hiroshi Kawauchi said. Also highly critical of the utility's information policies was Hiroyuki Fukano, who had become head of the government's Nuclear and Industrial Safety Agency (NISA) in August 2011. "Why don't they release all the information," asked Fukano in a newspaper interview. "There are problems with TEPCO's attitude toward providing information."[25] In the weeks after the accident, much of the public lost trust in the utility. In a telephone poll of more than 2000 randomly selected people outside of the three prefectures most affected by the disaster, 73% said they did not think information provided by TEPCO regarding Fukushima Dai-ichi was reliable.[26]

Communication in any disaster situation is challenging, particularly when so much infrastructure is damaged. Furthermore, no country in the world could have easily dealt with the kind of horrendous triple disaster that struck Japan. Everyone dealing with the disaster—from authorities to first responders to the public—faced extraordinary stresses. Still, as Japanese officials and others have made clear, numerous avoidable communication failures occurred in connection with the nuclear crisis. Many could have been prevented with better coordination, systematic advance planning, better messaging efforts, and more emphasis on keeping people informed.

The Public in Other Countries Also Had Information Needs

News of the crisis at Fukushima Dai-ichi spread quickly, and people in countries around the world became concerned about its implications for their health. One result was intense interest in potassium iodide (KI).

If exposure to radioactive iodine is suspected, authorities may recommend protective doses of potassium iodide. KI saturates the thyroid with regular (nonradioactive) iodine so that the gland does not take up the radioactive isotope. KI is only useful when people are at risk of exposure to radioiodine, and it only protects the thyroid. It is not a general "antiradiation" pill. Nevertheless, demand for KI in countries near and far skyrocketed. The Finnish Broadcasting Company reported that the Japanese nuclear accident had prompted an iodine rush. "In some cases, pharmacies reported that their supplies had sold out."[27] Pharmacies in the far eastern part of Russia experienced a similar run on potassium iodide and other medicines.[28]

In some places, where KI was unavailable, consumers purchased substances imagined to be good substitutes.[29] In China, shoppers mobbed stores to buy iodized salt. Long lines were reported in Shanghai, Beijing, and Hangzhou.[30] An exasperated Shanghai supermarket worker said: "We are entirely sold out of salt, and shoppers are now buying salt substitutes such as soy sauce, even though there is no connection."[31]

In fact, not only did people in other countries have no need for KI but also there were good reasons not to take it: the potential for allergic reactions, risks from overdosing, and risks from consuming iodine-based disinfectants or unproven herbal supplements.[32] The experience demonstrates the importance of providing correct information rapidly. It is essential that public health and medical agencies be proactive and highly visible. They must anticipate people's questions and be ready to immediately disseminate information that is responsive to public concerns.

Foreign Governments and International Bodies Wanted Information

Japanese officials provided updates to other nations and key international bodies about the crisis at Fukushima Dai-ichi. But there were several gaps in the effort. The most notable example involved the release of contaminated water to the sea.

Authorities made a decision to discharge a large volume of water with relatively low levels of radioactive contamination into the sea. The aim was to make room for the increasing quantities of more highly contaminated water that were accumulating at the nuclear facility.[33] However, Japanese authorities failed to notify other nations in the region that the step was being undertaken and criticism came quickly. This lapse was seen as a diplomatic blunder.[34]

In a report to the International Atomic Energy Agency (IAEA), Japanese authorities promised improved communications in the future.

> Notification to other countries including neighboring countries about the deliberate discharge of accumulated water of low-level radioactivity to the sea on April 4 was not satisfactory. This is a matter of sincere regret and every effort has been made to ensure sufficient communication with the international community and to reinforce the notification system.[35]

In the months after the sea discharge incident, Japanese officials worked to repair the damage and smooth international communication. After an IAEA site visit in late May and early June, the agency commented that the "Japanese Government, nuclear regulators and operators have been extremely open in sharing information and answering the many questions of the mission to assist the world in learning lessons to improve nuclear safety."[36] The controversy surrounding the sea discharge is a reminder that in an increasingly interconnected world, disaster communication must also include keeping other nations and the international community informed and updated.

Learning from Fukushima Dai-ichi

Thanks to the candor and comprehensiveness with which Japan has examined the handling of the 2011 Fukushima Dai-ichi accident, there are many lessons to be learned. The centrality of communication was underscored as was the importance of providing timely and accurate information. The accident also made clear that the range of information needs may be surprisingly broad and diverse. Populations requiring timely information can include

- People in affected areas
- Nearby or downwind communities
- The public in countries around the world
- Foreign governments and international bodies
- Special populations
- Hospitals, healthcare professionals, and emergency responders
- Members of the armed forces (including those involved in disaster assistance efforts)
- Evacuees
- Communities that may receive evacuees
- The media
- The nation as a whole

EFFECTIVE COMMUNICATION: PRINCIPLES FROM RESEARCH AND EXPERIENCE

In developing emergency communications, it is helpful to remember the wide variety of audiences noted above. From local to national to global, addressing these information needs can be crucial to an effective disaster and terrorism response.

One key principle is to *be proactive*. When an incident occurs, people immediately seek information about it and the actions needed to ensure health and safety.[37] If essential information is not available right away, rumors and spurious information may fill the void. But issuing emergency information immediately can be challenging. Disasters and terrorism events often occur without warning, "leaving agencies little or no time to develop effective communication strategies, informational materials, and emergency messages. In such a situation, events could easily outstrip communication efforts. . . ."[38]

On the basis of what was learned from the 2001 anthrax episode and other emergencies, the Centers for Disease Control and Prevention (CDC) and other agencies have increasingly adopted a more proactive approach to emergency communication. Part of this involves *pre-event message development*.[39] "In a nutshell, the idea is to carry out research on the concerns, information needs, and preferred information sources of key audiences; utilize the findings to prepare emergency messages and other materials; and carefully test them long before an event occurs."[40] Basic messages and materials are developed in advance, with incident-specific information added when an actual emergency occurs. By testing messages and message templates in advance, and by ensuring they are informed by research on people's information needs and preferences, agencies are able to release effective messages and information much more quickly.

Emergency information should be available for dissemination through a *variety of channels*, including print media, radio, television, and e-mail. Social media also play an increasingly important role in disaster communication.[41] Whether through sites such as Facebook, Twitter, MySpace, YouTube, LinkedIn, and Flickr, or through mobile phone text messages, forums, blogs/micro blogs, chat rooms, wikis, and bulletin boards, social media enable people to instantly share information. The use of social networks and online communities is especially important for reaching younger persons and households with children, since research suggests these groups make the greatest use of these modes of communication.[42]

Another key principle is that *information should be practical and actionable*. Information that is vague or too general may not prompt action, and telling people about a hazard without guidance about self-protection may result in passivity and fatalism.[43] "Messages should emphasize simple, practical steps and basic information about the threat."[44] Further, "action helps overcome feelings of hopelessness and helplessness."[45] Messages should include information about what to do and what *not* to do, as well as about who is at risk and who is *not* at risk. Action steps should also be feasible. "Authorities must let people know specific protective actions and explain why these actions will keep them safe."[46]

An additional key principle is that *information should be clear and easy to understand*. This means keeping communications relatively brief and free of long sentences, technical jargon, ambiguous wording, and complicated terminology. Clear communication increases the likelihood that it will be broadly understood including by people with low literacy. Approximately 14% of US adults aged 16 and older have "below basic" prose literacy and 12% have "below basic" document literacy. An example of a basic document literacy activity is using a television guide to find out what programs are on at a specific time. The same research suggests that 22% of the population is "below basic" on numerical literacy. An example of a basic numerical literacy activity is comparing the ticket prices for two events. Thus, simple and clear language will increase the likelihood that more people will be successfully reached with critical information.[47]

A further reason for the clearest possible messages is that in high-stress situations, everyone's ability to process information is significantly reduced. Risk communication expert Vincent Covello explains that "when people are stressed or concerned, they typically … have difficulty hearing, understanding, and remembering information." They "typically process information at four grades below their education level."[48] Reading level software and pretesting can help identify and remove complex, ambiguous, or unclear terms and concepts.

Another key principle is to ensure that communications are *credible*. Even if information can be disseminated rapidly, the effort may fail if the communications are not seen as accurate and believable. In crafting messages, it is extremely important to consider who is seen as having the requisite expertise and who the public trusts to provide accurate, honest information.[49]

Research suggests that at the federal level, health agencies are viewed by the public as the most credible sources of information in terrorism situations. This was a key conclusion of the Pre-Event Message Development Project, the largest focus group and interview study of terrorism risk communication issues to date.[50–52] Survey research has provided additional evidence that on terrorism issues, the public has higher levels of trust in federal agencies associated with health issues or perceived as having health expertise.[53]

Finally, another key principle is to *be consistent*. When a disaster or terrorist incident occurs, people look to multiple sources for information. They compare the various sources "to validate the veracity and accuracy" of the information.[54] If communications are inconsistent, "the public will lose trust in the response officials and begin to question every recommendation."[55]

An incident in London illustrates the value and importance of consistent, effective messaging. In 2006, a former officer in Russia's Federal Security Service who had been granted asylum by Britain was poisoned with radioactive polonium. The man, Alexander Litvinenko, had been a critic of Russian officials and policies. He subsequently died from the poisoning. Authorities found small traces of radioactive polonium at a sushi bar and hotel where the poisoning occurred, as well as in other locations around the city.

Given that radioactive materials were involved, and that media coverage was extensive, there was the potential for great public concern. But a survey of Londoners found that only about 11% believed their health to be at risk. One factor for the relatively low level of concern was that people viewed the incident as related to spying and espionage against a specific individual rather than an attack on the general public. But another reason was that health agencies had undertaken an extensive and consistent public messaging campaign to let people know who was potentially at risk and who was not.[56–58]

The principles identified above do not constitute a comprehensive list of everything important in emergency message and information development. But they do provide a foundation for preparing effective communications for disaster and terrorism incidents.

SPECIAL CONSIDERATIONS AND CHALLENGES

Along with the guiding principles highlighted above, it is useful to be aware of special considerations and challenges in developing messages and informational materials.

A Multilingual Society Requires a Multilingual Effort

According to the US Census Bureau, more than 47 million people aged 5 and above speak a language other than English at home.[59] The languages include Spanish (28.1 million), Chinese (2 million), French (1.6 million), German (1.4 million), Tagalog (1.2 million), Vietnamese (1 million), and Italian (1 million). About 55% of those speaking a language other than English report also speaking English very well. But that means millions do not. Thus, it is essential that messages and informational materials be available in a variety of languages.

Reaching Some Populations May Require Extra Effort

Not all groups are equally likely to see or hear disaster information. A study by Taylor-Clark et al. examining the 2005 Hurricane Katrina disaster found that employment and social connectedness of individuals correlated with their having heard the publicly issued advisories.

> People who were not employed at all were significantly less likely than were those who were employed full-time before the storm to have heard the evacuation orders. People who do not have family and friends, that is, had no social networks to rely on, were significantly less likely than those who were part of a network to have heard the evacuation orders.[60]

Thus, in providing critical information to the public, extra efforts and outreach may be required to communicate effectively with harder-to-reach subpopulations.

Parents and Younger Children Require Special Attention

In any disaster situation, parents and young children warrant special attention in communication and messaging. Children may not have the knowledge, coping skills, or emotional maturity to grasp the situation.[44,61] "They may be frightened by what they do not understand, and their misperceptions may lead to inaccurate interpretations and attributions."[62] Having age-appropriate explanatory materials such as coloring books and audiovisual resources could be helpful. So, too, can guides that help parents discuss a disaster or terrorist incident with their children.[63–65]

Pregnant or Breastfeeding Women Have Specific Information Needs

Pregnant and breastfeeding women comprise a special subgroup whose needs during a disaster or public health emergency are distinctive.[66] These women might have particular concerns about risks to the developing fetus or infant from a disaster or terrorism situation. Issues could include the risk of exposure to chemical, radioactive, or disease agents and whether breastfeeding should be continued. Additional concerns might be about the safety of food and water supplies and side effects of protective actions and medications.[67,68] Informational materials about these matters should be available, and authorities should be prepared to explain processes such as decontamination and population screening.

CRISIS AND EMERGENCY RISK COMMUNICATION RESOURCES

As the preceding sections make clear, communication and information issues in disasters and terrorism incidents are both crucial and challenging. Fortunately, many excellent tools and guides are available to make emergency communication more effective.

Several components of the US Department of Health and Human Services (DHHS) have developed practical materials on disaster and emergency communication. The Substance Abuse and Mental Health Services Administration has published *Communicating in a Crisis: Risk Communication Guidelines for Public Officials*. This handbook is an easy-to-use pocket guide on the basic skills and techniques needed for clear, effective communications, information dissemination, and message delivery. It includes discussions about emergency communication fundamentals, common myths, and communicating complex scientific and technical information.

The CDC has a variety of emergency communication resources, many of which can be found through the CDC's Emergency Preparedness and Response website (www.bt.cdc.gov). Here, one can access a series of four books on crisis and emergency risk communication (CERC): *Basic CERC, By Leaders for Leaders, CERC: Pandemic Influenza*, and *Pandemic Influenza Storybook*. The site also provides links to crisis and emergency risk communication Quick Guides and Wallet Cards in English and Spanish.

The CDC also offers fact sheets for disasters and terrorism situations. Some of the content is linked to translations in Chinese, Tagalog (Filipino), French, German, Haitian Creole, Italian, Korean, Portuguese, Russian, Spanish, or Vietnamese. The website also includes a DHHS toolkit of public health emergency text messages. The toolkit offers an array of pre-prepared cell phone text messages advising people how to protect their health after a disaster. The messages, which are intended to support state and local emergency managers in disaster response, cover everything from wound care and tetanus prevention to how to sanitize food cans after a disaster.

Another important resource is the First Hours web portal developed by the CDC and the Department of Health and Human Services' Office of Public Affairs. The portal is designed to assist local and state agencies in providing initial public messaging and information about new and emerging threats. In addition to a message template for the first minutes of all emergencies, the site includes sample messages, scripts, sound bites, and other communication resources for five broad categories of threat: bioterrorism, chemical agents, radiological emergencies, nuclear emergencies, and suicide bombs.

The Federal Emergency Management Agency (FEMA) has prepared brief guidelines on what steps people can take to protect themselves during various types of disaster or emergency (www.fema.gov/hazard/types.shtm). The Department of Homeland Security, working with the National Academy of Engineering and the Radio-Television Digital News Foundation, has developed *Communicating in a Crisis: News and Terrorism*. The project provides journalists, news managers, public information officers, and others with materials about public communication issues related to terrorism.

Finally, another excellent resource is the National Public Health Information Coalition (NPHIC), which is the network of public health communicators and information officers in the country's states and territories. The coalition maintains an online library of communication and messaging materials. They cover a variety of emergency subjects including carbon monoxide poisoning, extreme heat, severe weather, hurricanes, oil spills, radiation, West Nile virus, and pandemic influenza (www.nphic.org). The organization also hosts a discussion board on communication issues, holds an annual conference where new developments are discussed, and publishes a bimonthly newsletter.

Many other organizations, including agencies at local and state levels, also offer useful materials. But the above-noted resources provide a good starting point in developing more effective communications for emergency situations.

CONCLUSION

The case studies in the previous two chapters are about terrorist incidents, though of sharply different character. The London attacks in 2005 involved bombings of focused targets and the consequences were immediate and obvious. By contrast, in 2001, weeks passed before anyone realized that the United States was under biological attack. At that point, anthrax spores had already contaminated buildings and caused sickness and death.

Unlike those incidents, the Chernobyl, Bhopal, and Fukushima Dai-ichi disasters recounted in this chapter were not deliberately inflicted. But the unfolding of all these events, whatever their origin, was heavily influenced by communication issues. The centrality of communication is apparent as well in the following chapters that explore the roles of some of the principal responder groups.

NOTES

1. *Chernobyl's Legacy: Health, Environmental and Socioeconomic Impacts and Recommendations to the Governments of Belarus, the Russian Federation and Ukraine.* The Chernobyl Forum: 2003–2005. 2nd revised version. Vienna, Austria: International Atomic Energy Agency; 2006.

2. Ibid.

3. United Nations Scientific Committee on the Effects of Atomic Radiation [UNSCEAR]. *Sources and Effects of Ionizing Radiation.* UNSCEAR 2008 Report to the General Assembly with Scientific Annexes. Volume II. Scientific Annexes C, D, and E. New York: United Nations; 2011.

4. Chernobyl's Legacy.

5. Health effects of the Chernobyl accident: An overview. World Health Organisation Fact Sheet No. 303, April 2006.

6. Lillibridge SR. Industrial disasters. In: Noji EK, editor. *Public Health Consequences of Disaster.* New York: Oxford University Press; 1997. pp 354–372.

7. Shrivastava P. Long-term recovery from the Bhopal crisis. In: Mitchell JK, editor. *The Long Road to Recovery: Community Responses to Industrial Disaster.* Tokyo: United Nations University Press; 1996. pp 121–147.

8. Mishra PK, Samarth RM, Pathak N, Jain SK, Banerjee S, Maudar KK. Bhopal gas tragedy: Review of clinical and experimental findings after 25 years. Int J Occup Med Environ Health 2009;22(3):193–202.

9. Lillibridge.

10. Shrivastava.

11. World Health Organisation [WHO]. *Manual for the Public Health Management of Chemical Incidents.* Geneva: WHO; 2009.

12. Wray RJ, Becker SM, Henderson N, et al. Communicating with the public about emerging health threats: Lessons from the CDC-ASPH Pre-Event Message Development Project. Am J Public Health 2008;98(12):2214–2222.

13. Becker SM. Learning from the 2011 Great East Japan disaster: Insights from a special radiological emergency assistance mission. Biosecur Bioterror 2011;9(4):394–404.

14. Ibid.

15. Policy Planning and Communication Division, Inspection and Safety Division, Department of Food Safety, Ministry of Health, Labor and Welfare, Government of Japan. Issuance of instruction to restrict distribution of foods concerned, in relation to the accident at Fukushima nuclear power plant. Press release, March 21, 2011.

16. Nuclear Emergency Response Headquarters, Government of Japan. Report of the Japanese Government to the IAEA Ministerial Conference on Nuclear Safety—Accident at TEPCO's Fukushima Nuclear Power Stations, June 2011.

17. Ibid.

18. Ibid.

19. Onishi N, Fackler M. Japan held nuclear data, leaving evacuees in peril. The New York Times, August 8, 2011.

20. Nuclear crisis: How it happened/Government radiation data disclosure—Too little, too late. Yomiuri Shimbun, June 11, 2011.

21. Nuclear Emergency Response Headquarters, Government of Japan, June 2011.

22. Investigation Committee on the Accidents at Fukushima Nuclear Power Stations of Tokyo Electric Power Company. Executive Summary of the Interim Report. Tokyo, December 26, 2011.

23. TEPCO. White smoke around the Fukushima Daiichi Nuclear Power Station Unit 1. Tokyo Electric Power Company. Press release, March 12, 2011.

24. TEPCO submits more redacted Fukushima nuke plant manuals to Diet Committee. Mainichi Japan, September 13, 2011.

25. Head of nuclear watchdog criticizes TEPCO over blacked-out documents. Mainichi Japan, September 17, 2011.

26. Survey: 73 percent of Japanese do not trust TEPCO. Asahi Shimbun AJW, May 27, 2011.

27. Finns seek radiation protection with iodine pills. YLE Uutiset News, March 12, 2011. Updated March 14, 2011.

28. Schwirtz M. Fear of fallout from Japan spreads in Russia. New York Times, March 17, 2011.

29. Becker SM. 2011.

30. Pierson D. Japan radiation fears spark panic salt-buying in China. Los Angeles Times, March 18, 2011.

31. Rayner G. 2011. Japan nuclear fears prompt panic-buying around the world. The Telegraph, March 17, 2011.

32. Becker SM. 2011.

33. Agency didn't think to tell neighboring countries radioactive water was released into sea. Mainichi Japan, August 18, 2011.

34. Gov't, TEPCO should release more information on radiation contamination. Mainichi Japan, April 7, 2011.

35. Nuclear Emergency Response Headquarters, Government of Japan, June 2011.

36. International Atomic Energy Agency [IAEA]. International fact finding mission of the nuclear accident following the Great East Japan Earthquake and Tsunami, 24 May–1 June 2011. Preliminary Summary. Vienna: IAEA.

37. Wray RJ, Becker SM, Henderson N.

38. Becker SM. Social, psychological and communication impacts of an agroterrorism attack. In: Voeller JG, editor. *Wiley Handbook of Science and Technology for Homeland Security*. New York: John Wiley & Sons; 2008.

39. Vanderford M. Breaking new ground in WMD risk communication: The Pre-Event Message Development Project. Biosecur Bioterror 2004;2:193–194.

40. Becker SM. 2008.

41. Lindsay BR. *Social Media and Disasters: Current Uses, Future Options, and Policy Considerations*. Congressional Research Service, R41987. September 6, 2011.

42. American Red Cross. *Social Media in Disasters and Emergencies*. Online survey of 1,058 respondents representative of the US population aged 18 and older, conducted by Infogroup ORC on July 22–23, 2010. Report date: August 5, 2010.

43. Glik DC. Risk communication for public health emergencies. Annu Rev Public Health 2007; 28: 33–54.

44. Wray RJ, Becker SM, Henderson N.

45. Reynolds B. Crisis and emergency risk communication. Appl Biosaf 2005;10(1):47–56.

46. Wray RJ, Becker SM, Henderson N.

47. Kutner M, Greenberg E, Jin Y, Boyle B, Hsu Y, Dunleavy E. *Literacy in Everyday Life: Results from the 2003 National Assessment of Adult Literacy* [NCES 2007–480]. U.S. Department of Education. Washington, DC: National Center for Education Statistics; 2007.

48. Covello VT. Risk Communication: Warren K. Sinclair Keynote Lecture. NCRP Annual Meeting, March 8, 2010. Bethesda: National Council on Radiation Protection and Measurements.

49. Ronan KR, Johnston DM. *Promoting Community Resilience in Disasters: The Role for Schools, Youth, and Families.* New York: Springer; 2005.

50. Wray RJ, Becker SM, Henderson N.

51. Vanderford M.

52. Becker SM, 2004.

53. Kano M, Wood MM, Mileti DS, Bourque LB. *Public Response to Terrorism: Findings from the National Survey of Disaster Experiences and Preparedness.* Los Angeles: Regents of the University of California; 2008.

54. Wray RJ, Becker SM, Henderson N.

55. Reynolds B. *Crisis and Emergency Risk Communication: By Leaders for Leaders.* Atlanta, GA: US Department of Health and Human Services (HHS) in partnership with the Centers for Disease Control and Prevention (CDC) Public Health Practice Program Office, CDC Office of Communication (OC), and Office of the Director (OD); 2004.

56. Becker SM. Communicating risk to the public after radiological incidents. BMJ 2007;335(7630): 1106–1107.

57. Becker SM. Preparing for terrorism involving radioactive materials: Three lessons from recent experience and research. J Appl Secur Res 2008;4(1):9–20.

58. Rubin GJ, Page L, Morgan O, et al. Public information needs after the poisoning of Alexander Litvinenko with polonium-210 in London: Cross sectional telephone survey and qualitative analysis. BMJ. 2007; 335(7630):1143–1151.

59. Shin HB, Bruno R. Language use and English-speaking ability: 2000. Census 2000 Brief. Publication C2KBR-29. U.S. Department of Commerce, Economics and Statistics Administration, U.S. Census Bureau, October 2003.

60. Taylor-Clark KA, Viswanath K, Blendon RJ. Communication inequalities during public health disasters: Katrina's wake. Health Commun 2010;25:221–229.

61. Gurwitch RH, Kees M, Becker SM, et al. When disaster strikes: Responding to the needs of children. Prehosp Disaster Med 2004;19(1):21–28.

62. Pfefferbaum B, North CS. Children and families in the context of disasters: Implications for preparedness and response. Fam Psychol 2008;24(2):6–10.

63. Gurwitch RH, Kees M, Becker SM.

64. Becker SM. Psychosocial assistance after environmental accidents: A policy perspective. Environ Health Perspect 1997;105(S6):1557–1563.

65. Gielen AC, Borzekowski D, Rimal R, Kumar A. Evaluating and Creating Fire and Life Safety Materials: A Guide for the Fire Service. Produced by the Johns Hopkins Center for Injury Research and Policy for the National Fire Protection Association, December 2010.

66. Callaghan WM, Rasmussen SA, Jamieson DJ, et al. Health concerns of women and infants in times of natural disasters: Lessons learned from Hurricane Katrina. Matern Child Health J 2007;11:307–311.

67. Ibid.

68. Quinn D, Lavigne SV, Chambers C, et al. Addressing concerns of pregnant and lactating women after the 2005 hurricanes: The OTIS response. MCN Am J Matern Child Nurs 2008;33(4): 235–241.

PART II

HEALTHCARE PROFESSIONALS

4

THE ROLE OF THE EMERGENCY PHYSICIAN

Emily G. Kidd, Donald H. Jenkins, and Craig A. Manifold

HURRICANE KATRINA

In the early morning hours of September 2, 2005, several buses pulled into the parking lot of the Houston Astrodome. These buses contained men, women, children, and elderly people who had been sitting on a bridge in New Orleans for several days without food, water, or protection from the heat and elements. Most of them were simply tired, hungry, dirty, and scared, but there were some who arrived in Houston quite ill. A few were dead. No one in Houston, San Antonio, Dallas, or any of the other cities in Texas who had agreed to take in evacuees from New Orleans after Hurricane Katrina could have ever predicted the massive influx of people that occurred that day and in subsequent days. Over 400 buses flowed into the Astrodome parking lot on that first day alone, containing thousands of people looking for showers, water, clothes, shelter, food, and medical care.

The emergency medical community in Houston and other cities had gotten a tiny taste of what would occur that day on the evening before, when the National Disaster Medical System (NDMS) was activated to fly in some of the displaced hospital patients from New Orleans. Emergency physicians in local hospitals saw patients start arriving from Ellington Field via local emergency medical services (EMS) units on the evening of September 1, 2005. Many of these patients had decompensated medically from their stress over the previous days, and after being in the heat, vibration, and noise of the bellies of C-130 planes for several hours. Upon their arrival at the airfield in Houston, they were triaged by emergency physicians and other personnel in one of the hangars, and then dispensed to local hospitals for stabilization in the emergency departments and admission (Figure 4.1). The medical system handled the incoming patients in this fashion through the night, but the arrival of the buses the next day quickly overwhelmed the medical community, from prehospital EMS, to emergency departments, to the daily operations of huge hospital systems.

Local Planning for Terror and Disaster: From Bioterrorism to Earthquakes, First Edition.
Edited by Leonard A. Cole and Nancy D. Connell.
© 2012 John Wiley & Sons, Inc. Published 2012 by John Wiley & Sons, Inc.

FIGURE 4.1   EMS personnel from multiple services around the Houston area await C-130 planes filled with patients from New Orleans after Hurricane Katrina, September 2005. (Photo courtesy of David Almaguer, Assistant Chief, Houston Fire Department.) (For a color version of this figure, see the color plate section.)

The Houston medical community was forced to quickly devise an alternate plan to care for the incoming evacuees without overwhelming the local emergency departments and hospitals. The Harris County Hospital District worked with local hospitals and public health officials to set up a field hospital at the Astrodome/Reliant Center complex. Within just a few hours, a system was in place. People on the incoming buses were triaged and rapid initial medical evaluations were made by firefighters, paramedics, nurses, and emergency physicians from the Houston Fire Department.

Patients with emergent conditions were sent to nearby emergency departments via EMS units that were staged at the triage point. Those patients without immediately life-threatening problems were given initial treatment at the triage sites by the firefighters and paramedics, and also by local emergency physicians and nurses who had come to the Astrodome to assist. These patients were then sent by ambulance to the other side of the complex where a field hospital had been set up, complete with waiting area, treatment rooms, lab, pharmacy, and areas where patients who might traditionally be admitted to a hospital could be treated for longer periods of time with interventions such as intravenous fluids, antibiotics, nebulizer treatments, and monitoring. The next day, a second field hospital was set up in Houston's convention center by the University of Texas Health Science Center at Houston. This system resulted in thousands of people receiving needed medical evaluation and treatment while preserving the local emergency departments and hospitals for more serious medical emergencies and ongoing daily care of Houstonians.

In most disaster situations, emergency medicine physicians are the initial physician-level medical providers that patients will see. They must be prepared to triage and treat a multitude of disaster victims within a short period of time as well as manage prolonged increased patient volumes for months in some cases.[1] Of course, every emergency medicine physician should be adequately trained and possess the knowledge and skills to provide the standard of care in emergency medicine to their patients during any routine day or night. But what happens when the emergency physician is faced with a sudden surge of multiple patients with minor injuries from a bus accident that overwhelms the emergency department? What if patients arrive in the ER displaying signs and symptoms of exposure to a biological agent or chemical nerve agent? What if the hospital is hit by an earthquake or tornado (such as in Joplin, Missouri, in 2011) and loses power or its entire infrastructure? Or what if the physician's area has a catastrophic surge of patients that cannot be adequately cared for in the small footprint of one, or even multiple, emergency departments and hospitals, as in Hurricane Katrina in 2005?

Emergency physicians should be knowledgeable and proficient not only in emergency medical care based in emergency departments with intact resources and infrastructure but also in care specific to potential disaster situations, including emergency medical triage of multiple patients; both prehospital care and care in austere environments; their hospital's disaster and medical surge plan; their responsibilities when responding to the scene of a disaster; and local, regional, state, and federal emergency management and disaster plans and procedures. Training in disaster preparedness and response should begin during residency[2] and be a continuous process throughout an emergency physician's career.

FUNDAMENTALS OF EMERGENCY TRIAGE BY THE EMERGENCY PHYSICIAN

The term "triage" applies to a wide spectrum of processes in the delivery of medical care. Basically, it is the process by which casualties are evaluated and classified to facilitate an appropriate distribution of medical care. The process is driven by the principle of accomplishing the greatest good for the greatest number of people. The discussion that follows outlines some of the core tenets of triage and how they are applied to the unique situations that arise in the delivery of disaster medical care.

The process of triage is complex, requiring a working knowledge of multiple factors and applying them appropriately. These factors include, but are not limited to, current casualty flow, expected casualty flow, current resources available, and the ability to replenish these resources in a timely fashion. When resources are abundant, which commonly is the case in civilian trauma centers, the maximal effort can be expended on each patient during normal, day-to-day operations. In this situation the sorting criterion becomes "sickest patient first." Even if the chances of a good clinical outcome are minimal, the criterion is justified because no other patients are put at risk.

However, when a maximal effort cannot be applied to each additional patient, as in a disaster or mass-casualty incident, the decision process evolves to that of triage. Casualties with catastrophic injuries that limit the chance of survival are passed over until those with life-threatening but "reversible" injuries are treated. This concept was known even when lifesaving surgical procedures were rare. In the mid-nineteenth century, John Wilson in "Outlines of Naval Surgery" said that lifesaving surgery could only be made available to those most in need if treatment was withheld from those whose injuries were likely to prove fatal, and deferred for those whose injuries were slight.[3] These principles have become universal and must be applied whenever the demand for medical care outstrips the supply.

The unique aspects of disaster triage stem from situations in which this medical care is delivered. Deployed medical teams and equipment must be mobile enough to adjust to changes in conditions. This restriction means all medical care is delivered from limited resources. The imminent danger to patients and providers must be weighed in the decision of where care is to be provided. Situations may arise where standard medical triage decisions based on the greatest good are replaced by other considerations. This would be the case, for example, when resource allocation is redirected to casualties with injuries that could be treated and then returned to critical duties. A variety of concepts and technologies that are currently being applied to triage and early resuscitative interventions will be discussed below.

The Triage Officer

Who should perform the triage function? Historically, the military advocated the use of the most experienced surgeon to perform this function.[4] In some scenarios, the US Air Force designates the senior Flight Surgeon to perform these duties. Dental officers, obstetricians, and senior medical corpsmen have also been utilized in this capacity.[5] The emergency medicine physician is an ideal triage officer.[6,7] Extensively trained to evaluate, resuscitate, and manage patients, emergency physicians have the knowledge base and procedural competence to manage the disaster casualty initially. Emergency medicine (EM) specialists frequently make triage decisions in civilian multiple casualty situations. These EM specialists are becoming the norm in military field medical systems.[8,9]

In Desert Storm, during the 1991 Gulf War, emergency physicians were utilized in the triage officer role. One military field trauma center had 461 coalition and enemy personnel triaged as casualties. Only 7% of enemy forces and 2% of coalition casualties required re-triage. Twelve of 302 (7%) enemy forces were re-triaged from a delayed to minimal category representing "over-triage." Three (2%) of coalition forces were re-triaged from delayed to immediate. Outcomes of these three patients were not reported.[10] If an emergency physician is available, she should be assigned the triage role.

Personnel performing triage must be competently trained in making such decisions. While there is no substitute for appropriate training and experience, these features must also be coupled with sound clinical judgment.[11] Psychological factors may affect this decision-making process. The responsible physician must not allow emotion to influence his or her triage decisions at the critical time of action. Sound medical judgment and appropriate utilization of resources must remain the basis for triage decisions. The triage physician will be in the position of surrogate decision-maker for the patient. This singular responsibility is uncommon in day-to-day civilian operations. Now, a patient who lacks decision-making capacity because of injury could be subject to the biases of the triage and treating physician. Providers entrusted with this role must be wary of untoward preconceptions as they make difficult decisions with minimal additional guidance.

Triage Techniques

Many triage systems exist and the practitioner should be familiar with the triage techniques practiced in her healthcare system and region. A few principles are key to understanding and improving triage abilities. The practitioner should attempt to distinguish among three classes of patients: those who are likely to survive whether or not they receive care, those who probably will die no matter what services they receive, and those who will benefit from interventions that can be performed with the resources available (Figure 4.2).

FIGURE 4.2 EMS personnel, nurses, and emergency physicians care for ill and injured Hurricane Katrina victims as they arrive on buses at the Astrodome, September 2005. (Photo courtesy of Diana J. Rodriguez, Houston Fire Department.) (For a color version of this figure, see the color plate section.)

Geographic triage is helpful in events where patients are concentrated in an area. This method consists of requesting all patients who need assistance to move to a designated area for additional help. The process self-selects patients who are mentating and can ambulate, reflecting a relatively minor injury. Next, the triage physician should ask all people who still need assistance to raise their hand or yell out. This method also selects patients who are mentating but may benefit from intervention. The remaining patients may not benefit from any action.

Reverse triage is a method used in the military setting whereby the least injured would be treated first in order to return soldiers to duty quickly. Advanced or secondary triage may take place later at a facility and include ancillary data such as lab tests or radiographs to assist in decision-making.

For initial processing, many systems use a predefined protocol that incorporates limited clinical presentations and vital signs data. A color coded or tagging mechanism is then used for prioritizing transport decisions. EMS personnel in the United States commonly use the Simple Triage and Rapid Treatment (START) system. START incorporates a quick assessment of a patient's respiratory status, perfusion, and mentation. This method requires minimal training and can be quickly and easily incorporated into training programs. Additional systems are available and widely used around the globe. There is minimal outcome data to support the use of any one process over another. The important point is that the emergency physician be knowledgeable about the system or protocol in his or her response area, along with the techniques applicable to potential disasters in that area.

The optimization of initial provider performance of lifesaving maneuvers (airway interventions, thoracic decompression, limb stabilization, and immobilization) will save many lives. Recognizing mental status as a sensitive indicator of tissue perfusion, and classification of non-responders and transient responders to initial therapy, will identify most patients requiring critical care resources. Transport prioritization can then be established.[12]

Triage, a dynamic process, is not isolated to initial evaluation. Critical decision points occur through the time of injury to stabilization at a definitive care facility. Identifying patients who require excess resources in order to salvage life or limb can be difficult. During triage, physiologic and biochemical data available at the time of decision may be limited. Available resources may be the determining factor in the decision-making process. This is contrary to daily practice where it is relatively rare to not have enough consumable goods to treat all victims. Plain film radiography and computed tomography, available at every civilian level 1 trauma center, may not be available at deployed disaster facilities. Providers must rely on physical examination and changing hemodynamic parameters for triage decision-making.

Identifying non-survivable wounds, or injuries with low likelihood of survival, may not be easy. Because of limitations on the quantities of crystalloid and blood products available, these items should only be used first on patients who clearly would benefit. Conservation limits can be determined early during a mass casualty incident and then communicated to the entire staff. Establishing prior standards will decrease the need for instantaneous decisions at the time of operative intervention.

Triage is dynamic and should take place along many points of patient care and transport: initial triage at the scene, prior to transport, and at the receiving facility where determinations are made about the application of surgical resources or ventilator rationing. Healthcare facility triage has its own set of issues and guidance regarding palliative care and resource availability. For example, the State of Utah's Department of Health includes in its Pandemic Influenza Triage Guidelines categories of patients that are lowest on the priority list for ventilator and ICU support. They include patients with "do not resuscitate" orders, advanced dementia, organ failure, incurable malignancy, fatal chromosomal disease, severe burns, and other conditions with very low probabilities of survival.

Current technological advances, such as ultrasound, lactate measurement, capnography, and computer-aided decision algorithms, may augment current triage practices. However, nothing will replace basic assessment skills and experience of the triage officer. Many applications available in current practice are beneficial in the triage setting. These technologies must remain reliable and inexpensive to be utilized in the mass casualty setting.

Triage indisputably saves lives. Still, rules and critical pathways regarding triage can be difficult to articulate, or to demonstrate with controlled experimental data. Relating anecdotal experiences remains the most common form of education regarding triage principles. Optimal triage requires knowledgeable personnel with the training, experience, and resolve to make difficult decisions based on limited data. Education and practice in realistic environments with changing resources are the most effective means of learning these skills.

PREHOSPITAL MEDICAL CARE AND THE EMERGENCY PHYSICIAN

Most areas in the United States have some form of EMS with personnel trained to respond to emergency calls in emergency vehicles, provide some level of emergency medical care, and transport patients to an emergency department for physician-level care (Figure 4.3). The level of care in EMS can range from first responders who are trained only in basic first aid

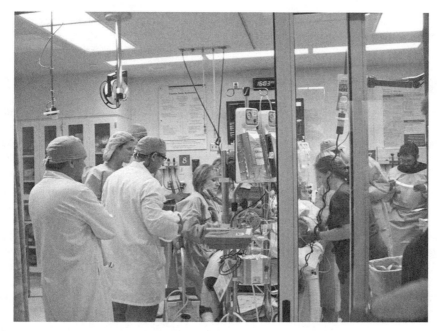

FIGURE 4.3 A Code Yellow (multiple patients) response at Mayo Clinic. (Photo courtesy of Donald Jenkins, MD.) (For a color version of this figure, see the color plate section.)

and CPR, to advanced level paramedics and nurses who perform advanced airway procedures, deliver cardiac and other medications, obtain intravenous and intraosseous access for medications and fluids, and perform a myriad of other medical procedures prior to arriving at a hospital.

During a disaster, the emergency physician will likely be interacting closely with the EMS personnel in his or her area. Thus, the physician should be comfortable with the level of care provided to the patients in advance of his or another doctor's seeing them. In preparing themselves for a disaster, emergency physicians should be thoroughly familiar with all EMS in their area. Emergency physicians should have knowledge of what areas each EMS system covers, what the capabilities of the EMS personnel are for each system, who provides the physician medical direction for each service (this is usually a local physician), and what each EMS system's disaster response would entail.[13]

As a rule, emergency physicians, nurses, and other emergency medical personnel should not self-dispatch to the scene of a disaster, and should only respond to the scene if adequately trained *and requested*.[14,15] At the same time, if called upon, they should be able to perform credibly in prehospital settings. They should be knowledgeable enough in prehospital medicine to function in an environment without the traditional amenities available in the emergency department. During the Hurricane Katrina experience, emergency physicians were forced to treat patients for a period of time in prehospital settings with only EMS supplies and without the traditional "comforts" of a laboratory, radiology capabilities, and inpatient services. This scenario might also play out in a hospital emergency department itself if power or other infrastructure is interrupted. In preparation for responding to a disaster, emergency physicians should consider periodically riding out with their local EMS to keep familiar with the environment and the services' protocols and capabilities.

EMERGENCY MEDICAL CARE SPECIFIC TO DISASTER SITUATIONS

Emergency physicians will be among the frontline soldiers in detecting biological threats or outbreaks. All physicians, but especially emergency physicians, must remain alert and vigilant for unusual symptoms or clusters of symptoms that might indicate a biological emergency (fevers, GI symptoms, respiratory symptoms, rashes). It can be challenging for individual emergency physicians to notice patterns, given their patterns of shift work and the non-specificity of many symptoms of biological agents. For this reason, the aggregation of real-time reporting and the sharing of information from several physicians may be needed to detect unusual patterns.[16]

In the event of a biological outbreak, whether from an attack or natural causes, all healthcare providers must protect themselves with suitable personal protective equipment (PPE). Determining the appropriate PPE requires monitoring public health guidance (which could change frequently) from local or state public health officials, or the Centers for Disease Control and Prevention (CDC). Standard blood-borne pathogen precautions always apply: isolation of all bodily fluids of affected patients, mask and eye protection, appropriate hand washing. Other precautions may vary by biological agent,[17] and might include avoidance of droplet exposures (private patient rooms, masks), contact precautions (private patient rooms, gloves and gown), and airborne precautions (negative pressure isolation rooms and appropriate respiratory protection such as N95 [particulate respirator] masks or SCBA [self-contained breathing apparatus] masks).[18]

Emergency physicians should keep abreast in their knowledge about presentation, prevention, prophylaxis, and treatment of different biological agents. In the event of an outbreak or pandemic, they must stay tuned in for public health guidance. The need was starkly demonstrated during the H1N1 pandemic in 2009, when public health guidance on PPE, laboratory testing, prophylaxis, and treatment changed almost daily. Emergency physicians and others had to keep communication channels open continuously to remain current.

If faced with victims of a bombing or explosion, emergency physicians must be knowledgeable and proficient in the recognition and care of blast injuries. Primary blast injuries result from barotrauma to air-filled organs (middle ear, sinuses, lungs, GI tract) from the initial pressure changes of the blast. Such injury to the lung may require complex ventilatory, fluid, and support management.[19] Secondary blast injuries result from penetrating wounds caused by flying debris. They warrant suspicion of highly serious injury that would require emergent surgical intervention. Tertiary blast injuries include wounds caused by the patient being thrown by a blast wave; they are usually blunt injuries and fractures.

Finally, quaternary blast injuries encompass all other injuries related to a blast not otherwise described. These include burns, crush injuries, chemical inhalational injuries, and the like.[20] Emergency physicians should anticipate that after an explosion, hospitals could see an "upside down" triage, meaning that the less seriously injured will surge upon the hospitals closest to the blast site before the more seriously injured arrive. These less-injured individuals may have to be diverted to hospitals farther away from the blast in preparation for the more severely injured.[21,22]

Chemical disasters may be caused by industrial accidents, transportation accidents, or terrorist events. A comprehensive review of the health issues related to chemical exposures is not possible in this chapter, but emergency physicians should be familiar with the effects

FIGURE 4.4   Emergency department personnel decontaminate patients for an airport drill. (Photo courtesy of Donald Jenkins, MD.) (For a color version of this figure, see the color plate section.)

of various hazardous materials. They should understand what information should be sought from the first responders to help identify the causative chemical agent. Further, they should be aware of expected symptoms from exposure, which PPE should be utilized by emergency department staff, and which decontamination procedures should be employed (Figure 4.4).

Natural disasters come in various forms, which can present different challenges to emergency physicians. Earthquakes cause injuries ranging from minor cuts to crush wounds to burns from ensuing fires. Respiratory injuries from dust and debris may surface early after an earthquake, and last for days or longer. Crush injuries require specific recognition and therapy. The emergency physician should be proficient in the treatment of all these conditions.

Hurricanes can lead to a variety of medical and traumatic emergencies. The obvious problems, such as drowning and injuries from flying debris and collapsing buildings, may be aggravated by difficulties related to pre-storm evacuations and post-storm activities. In 2005, prior to the arrival of Hurricane Rita, much of Houston, Texas was evacuated. The mass movement of frail and elderly persons, and those with chronic medical conditions, posed huge challenges. Many of them had not made emergency plans and they lacked adequate hydration, food, and regular medications. Some experienced decompensation (deteriorated heart function) under these stressful conditions.

FIGURE 4.5 A massive wildfire destroys hundreds of homes and causes medical problems for firefighters and citizens in Bastrop, Texas, September 2011. (Photo courtesy of Emily Kidd, MD.) (For a color version of this figure, see the color plate section.)

Emergency physicians should also anticipate dealing with these and other problems in the post-storm phase. They could face issues related to clean up and recovery such as heat emergencies, carbon monoxide poisonings, traumatic injuries, even animal bites. Many of these problems might be seen during and after flood events as well.[23]

Wildfires can cause havoc on local emergency rooms. Burns and heat injuries are often seen during fires, and emergency physicians might have to treat firefighters and other burn victims (Figure 4.5). Also, likely to be seen are exacerbations of respiratory and cardiovascular illness, allergies, and complications from exposure to poisonous gases. These complications can develop in individuals miles from the actual fire. Like most other natural disasters, wildfires have a disproportionate impact on vulnerable populations.[24]

Finally, emergency physicians should be prepared to deal with delayed care issues from natural disasters for weeks to months after the event. These could range from infectious diseases caused by contaminated food, water, and mold to psychological trauma.[25]

EMERGENCY MANAGEMENT KNOWLEDGE, SKILLS, AND RESPONSIBILITIES

A system of incident management for disasters was first introduced after a series of catastrophic wildfires in California in the 1970s. Many people died as a result of deficiencies in communication and management among multiple responding agencies. After the attacks on the World Trade Center in 2001, the federal government improved its framework for disaster management throughout the country including at national, regional, state, and local levels. This framework was initially outlined in a document called the Federal

Response Plan, which was improved upon and, after September 11, renamed the National Response Plan. In 2008, it was further revised and retitled the National Response Framework (NRF).[26]

Historically, there has been confusion about "who is in charge" during a disaster. In most instances, the management of a disaster starts and ends at the local level, and is the responsibility of the highest elected official of a jurisdiction (mayor, county judge, parish president). These officials have the ultimate responsibility for planning for disasters in their jurisdiction. They commonly employ an emergency management coordinator, who should be an expert in the four phases of emergency response and incident management: mitigation, preparedness, response, and recovery. Emergency management directors and coordinators should first use available local and mutual aid resources to manage a disaster. Only if local resources are depleted should they turn to state and federal resources. It is imperative that emergency physicians understand their area's emergency management structure, and how medical care fits into that structure.

The Joint Commission, which is the accrediting organization for most hospitals in the United States, requires that hospitals have disaster and emergency management plans. Each plan should contain provisions for internal and external disasters. Internal disasters include in-hospital emergencies such as fires, power outages, active shooters, and disruption of hospital infrastructure. The protocols should indicate how every department should respond in such situations. External disasters can range from natural events like hurricanes and earthquakes to biological outbreaks, warfare, civil disorder, terrorist acts, and transportation accidents.[27] The disaster plans must provide both for bringing patients to the hospital as well as evacuation from the hospital.

These plans should also include the hospital's incident command system, assessment of hospital capabilities, role assignment to staff members, personnel recall, records maintenance, resupply, security, and public relations.[28,29] Other specifics on management of different disasters should be addressed as well, including the management of a surge of patients, designation of alternate care sites, evacuation of the hospital, and decontamination of patients prior to entering the facility.

A hospital disaster plan should include a hazard-vulnerability assessment for that facility. The overall plan should be periodically reviewed and tested with regular exercises that involve others in the responder community. Integrating a disaster plan and disaster drills with local fire, EMS, law enforcement, emergency management, public works, and governmental entities is essential. As the person who will be leading the management of disaster victims, the emergency physician should know precisely where he or she fits into the hospital's incident command structure. The emergency physician should be actively involved in all disaster planning and drills in which the hospital participates.[30]

Although emergency physicians should be able and willing to work at the scene of a disaster, they should not self-dispatch to a disaster scene. National Incident Management guidelines, along with the National Association of EMS Physicians (NAEMSP) and the American College of Emergency Physicians (ACEP) maintain that medical personnel should not respond to a disaster scene unless officially requested by the jurisdiction's established incident commanders. If requested to respond to a scene, they should fall under the command of the local EMS medical director and his or her position in the incident command system.[31]

In addition to the requirements already noted, emergency medical physicians and other personnel responding to a disaster should: assure that their primary institution is not in need of their services before they respond to the scene (meaning that the first responsibility of

FIGURE 4.6 Hospital Incident Command Center at Mayo Clinic. (Photo courtesy of Donald Jenkins, MD.) (For a color version of this figure, see the color plate section.)

the emergency physician is to his or her primary institution); understand the basics of the National Incident Management System (NIMS) and Incident Command Structure (ICS) and where the emergency physician falls in the ICS of the disaster scene; and understand what PPE may be required and how to properly use it.[32] (Figure 4.6)

Emergency medicine physicians must similarly be prepared to deal with a "secondary disaster" at their institution when unsolicited medical personnel, blood donors, and other well-meaning volunteers converge on the emergency department to help. Their presence in large numbers poses problems ranging from added demands on resources to dealing with injuries sustained because they did not understand or adhere to principals of personal safety.[33]

TRAINING AND EXERCISE FOR DISASTER PREPAREDNESS

As noted earlier, disaster preparedness training should start at the residency level for emergency physicians. While most emergency medicine residency programs in the United States recognize the importance of some form of disaster training, the actual content of a program may vary substantially from one institution to another. This disparity should be addressed by standardizing disaster curricula and training across emergency medicine residencies.[34] After formal training, continuing education on disaster preparedness by the emergency physician is crucial, as is active participation in regular exercises and drills.

The Joint Commission requires an annual exercise of each hospital's preparedness plan in the form of a community-wide disaster drill. Emergency physicians should be actively involved in the planning and implementation of these drills. The drills should include simulated patients, practice with the communications systems, and incorporate local EMS systems, law enforcement, emergency management, and volunteer agencies.[35,36] It is critical that each participant in the emergency response knows and performs his or her roles during the drills. Drills should focus not solely on patient treatment, but also encompass triage, patient tracking and documentation, decontamination, and inter-organizational agreements.

CONCLUSION

Emergency physicians play a central role in planning for, responding to, recovering from, and mitigating terrorist and disaster incidents. They are on the front line in physician-level care and therefore have a huge responsibility to be familiar with the medical care associated with these events. Further, they must understand their roles in emergency management and incident command structure, in dealing with other first responders, and in preparedness in their own facilities.

NOTES

1. Kaji AH, Waeckerle JF. Disaster medicine and the emergency medicine resident. Ann Emerg Med 2003;41(6):865–870.

2. Ibid.

3. Kennedy K, Aghababian RV, Gans L, et al. Triage: Techniques and applications and decision making. Ann Emerg Med 1996;28(2):136–144.

4. Bowen TE, Bellamy RF, editors. *"Triage" in Emergency War Surgery*. Washington, DC: U.S. Department of Defense; 1988. pp 181–192.

5. Ryan JM. The Falklands war-triage. Ann R Coll Surg Engl 1984;66:195–196.

6. Naggan L. Medical planning for disaster in Israel. Injury 1976;7(4):279–285.

7. Walsh DP, Lammeright GR, Devoll J. The effectiveness of the advanced trauma life support system in a mass casualty situation by non-trauma experienced physicians: Granada, 1983. J Emerg Med 1989;7:175–180.

8. Llewellyn CH. Triage: In austere environments and echeloned medical systems. World J Surg 1992;16:904–909.

9. Walsh, Lammeright, Devoll. 175–180.

10. Burkle FM, Orebaugh S, Barendese BR. Emergency medicine in the Persian Gulf War. Part I: Preparations for triage and combat casualty care. Ann Emerg Med 1994;23(4):742–47.

11. Kennedy, Aghababian, Gans. 136–144.

12. Forward Resuscitative Surgery Report. Prepared for the Joint Health Service Support Vision 2010. Department of Defense, 1997.

13. Asaeda G, Cherson A, Richmond N, *et al.* Unsolicited medical personnel volunteering at disaster scenes: A joint position paper from the National Association of EMS Physicians and the American College of Emergency Physicians. Prehosp Emerg Care 2003;7(1):147–148.

14. Ibid.

15. Kaji, Waeckerle. 865–870.

16. Cochrane D. Perspective of an emergency physician group as a data provider for syndromic surveillance. MMWR 2004;53(Suppl):209–214.

17. Centers for Disease Control and Prevention website: www.cdc.gov

18. Clements BW. *Disasters and Public Health: Planning and Response*. Oxford: Butterworth-Heinemann; 2009. pp 55–57, 67–80, 158–160, 202–203, 280–283.

19. Ibid.

20. Ibid.

21. CDC website: www.cdc.gov

22. Clements. pp 55–57, 67–80, 158–160, 202–203, 280–283.

23. Ibid.

24. Ibid.

25. Ibid.

26. National Response Framework, Federal Emergency Management Agency, U.S. Department of Homeland Security, Washington, DC, 2008.

27. The Joint Commission website: www.jointcommission.org

28. Goolsby CA, Brenner BE, editors. Introduction to disaster planning: The scope and nature of the problem. Medscape Reference, 2011. Available at http://emedicine.medscape.com/article/765495-overview, accessed June 17, 2012.

29. Koenig K. New standards in emergency management: Major changes in JCAHO requirements for disasters. ACEP Disaster Medicine Section Newsletter, Vol. 10, Issue 1, June 2001.

30. Goolsby, Brenner.

31. Asaeda G, Cherson A, Richmond N, et al. pp 147–148.

32. Ibid.

33. Kaji A, Waeckerle J. pp 865–70.

34. Kefalas S, Shalkham A. Resident education and training in disaster medicine. ACEP Disaster Medicine Section Newsletter, Vol. 15, Issue 1, January 2006.

35. Kaji, Waeckerle. pp 865–870.

36. The Joint Commission website: www.jointcommission.org

5

THE ROLE OF THE NURSE

Dian Dowling Evans, Samuel E. Shartar, and James W. Gordon

THE OLYMPIC PARK BOMBING

At 12:30 A.M., July 27, 1996, Olympic Village in Centennial Park, the designated "social center" for visitors to the games in Atlanta, was jammed with 10,000 people.[1] While young and old were partying and crowding around the music venue stage listening to the band *Jack Mac and the Heart Attack*, a man made his way to a bench near the stage. He placed his backpack under the bench and disappeared back into the crowd. After milling about for several minutes he began to head for the exit.

Eric Robert Rudolph, 29, had been planning this night for over a year. An antiabortion fanatic, he had studied the best way to build a small yet powerful bomb and where to place it for the greatest impact on those nearby. His objective was to kill, maim, and produce terror at the "event of the century in Atlanta." To achieve this goal he had taped together three separate pipe bombs that could be activated by a battery-powered timing mechanism. In his backpack, next to the bombs, he had placed a plastic container full of screws and nails to increase the damage.[2]

Near the exit, Rudolph stopped at a pay phone and placed a 911 call. In a calm voice he told the dispatcher, "There is a bomb in Centennial Park. You have 30 minutes." As he left the park he noticed several security officers, relieved none of them had glanced his way.

Five minutes earlier, at approximately 1:00 A.M., a security officer was approaching the front of the stage where an altercation had erupted between two young men. After breaking up the fight, he noticed an unattended backpack at the bench nearby and, per protocol, radioed for backup to investigate. Around the same time the band's manager also had noticed the suspicious backpack and called to alert security. Within minutes several officers moved into the area followed by a media cameraman who had been filming the security operation. As officers began to clear spectators from the area the backpack exploded. The detonation sprayed the nails, screws, and other fragments into the crowd for a distance of 100 feet in all directions. Moments later the cameraman collapsed with what would later

Local Planning for Terror and Disaster: From Bioterrorism to Earthquakes, First Edition.
Edited by Leonard A. Cole and Nancy D. Connell.
© 2012 John Wiley & Sons, Inc. Published 2012 by John Wiley & Sons, Inc.

be diagnosed as a massive myocardial infarct. One woman lay dead from a shrapnel wound to her head while others lay bleeding from body injuries.[3,4]

As people began pouring out of the area, firefighters and ambulance personnel moved in from nearby stations to tend to the victims. The most seriously injured were quickly triaged and transported to Grady Memorial Hospital, an affiliate of Emory University and the only level 1 trauma center in Atlanta. Although Grady was a mile from the blast, it had not been designated an official affiliated hospital for the Olympics. Nevertheless, its chiefs of medicine and surgery had been advisers to the Atlanta Olympics Planning Committee in discussions specific to disaster preparedness and response.[5] In addition, given Grady's proximity to the games, its physicians and nursing staff had developed hospital-wide disaster response plans, which included participation in department disaster drills.[6] In fact, 1 week before the start of the games, Grady physicians and nurses had joined a team of firefighters, emergency medical technicians (EMTs), and security personnel at Atlanta's Hartsfield International Airport in a mock mass disaster drill.[7]

Advance Preparations

Medical preparations for the Olympics had taken 5 years to complete.[8] The Centers for Disease Control and Prevention (CDC), in conjunction with medical assessment teams from the Army and the Federal Bureau of Investigation (FBI), established a center at the CDC to provide immediate emergency public health, toxicological, forensic and medical assistance in the event of a biological or chemical attack. Specialized teams from a variety of military and governmental agencies were also recruited to provide assistance in the event of a terrorist attack. Local hospital emergency departments were prepared as well in anticipation of the 2.2 million visitors including 35 heads of state and 10,000 athletes expected to attend the games. In addition, Public Health Service Disaster Medical Assistance Teams staffed by nurses, physicians, and EMTs were brought into Atlanta and stationed at a nearby Air Force base ready for helicopter deployment if further surge capacity was needed.[9]

Security operations for the Olympics were the largest in the history of any peacetime event in the United States. Thirty thousand security personnel recruited from the Bureau of Alcohol, Firearms and Tobacco, FBI, Georgia Bureau of Investigation, local police, military, and national guard, at a cost of 303 million dollars, had been assembled.[10] In the months preceding the games no terrorist threats targeting the Olympics had been made, and the greatest security threat was thought to be potential street gang violence.[11]

For weeks leading up to the games, medical and security personnel engaged in several mass casualty disaster drills including preparation for nuclear, chemical, and biological attacks. A mock kidnapping and an aircraft hijacking drill were also undertaken.[12] During the first week of the Olympics security personnel, working on heightened alert, seized approximately 100 unattended bags and searched them for hidden explosives. Dedication ceremonies at the Olympic Village dormitories on the Georgia Tech campus were interrupted when security noticed an unclaimed bag. The attendees were evacuated though fortunately, no explosives were found.[13]

The 21-acre Olympic Village in Centennial Park, where the blast occurred, had been open to the public. The area was filled with walkways, vendor stands for shopping, tented comfort stations, fountains, and included the music stage.[14] Despite the large number of security personnel assembled for the games, this particular area was meant to be accessible to visitors. Entry did not require a ticket or passage through a metal detector.[15] Fire department personnel traveling on golf carts and police and emergency medical personnel in ambulances were stationed in multiple locations for a rapid response in case of an emergency. Additional ambulances had been brought in from around the country to augment existing resources.[16] This increased manpower-enabled rescuers to triage 96 of the 111 victims of the blast within the first 30 minutes of the explosion. Patients were distributed to 11 area hospitals, though the largest numbers were taken to Grady.[17]

Initial Response Efforts

Within minutes after the blast, a call came in on the ambulance radio to the Grady Hospital Emergency Care Center (ECC): "Grady ECC, Grady ECC, this is Grady Unit 34, Grady go ahead." The triage charge nurse answered the radio expecting routine information. But then:

> Grady ECC this is not a drill, repeat, this is not a drill. There has been a large explosion at the Olympic Park in downtown Atlanta. There are multiple casualties and triage has started; will advise further as patient condition information is available.

The nurse immediately informed the chief attending physician that there had been an explosion nearby with several casualties. About the same time, a report of the bombing had also reached Grady's Chief of Medicine who immediately ordered implementation of the hospital disaster plan. Word then spread quickly in the ECC about the possibility of a mass casualty disaster event.

The 17 nurses, 13 physicians, and 1 physician assistant on duty were already busy with the usual Saturday night emergency department issues including the resuscitation of two critical patients. But they responded without a trace of panic, shouting, or other visible sign of anxiety.[18]

NURSES AND INTERDISCIPLINARY RESPONSE EFFORTS

Working collaboratively, Grady ECC nurses and physicians began to ready the triage area and trauma bays for the arrival of victims where they would be joined by in-house trauma surgeons. Charge nurses in different areas of the ECC began organizing the nursing staff and giving assignments in preparation for patient arrivals. One of the nurses went to the patient waiting room to announce that an emergency had occurred and that an unknown number of injured people would be arriving. Most of those waiting left to make room for the anticipated victims. Patients already in the ECC treatment area were rapidly triaged and those whose conditions were neither serious nor critical were directed to the waiting room to clear treatment space for blast victims. The majority of these patients soon voluntarily left the hospital to return at a later time.[19]

Procedures for hospital-wide phone calls were initiated as part of the disaster plan to recruit additional nurses and physicians to staff the operating rooms, intensive care, medical and surgical units, and the emergency department. Within the first hour, 126 additional nurses reported for duty. Intensive care nurses already on duty began to triage patients, moving those who were less critical to intermediate care units, to make way for blast victims. Nurses on the medical/surgical floors identified 81 patients who could be discharged safely to make room for disaster victims if necessary. Operating room nurses and support staff readied rooms as additional surgeons and surgical residents reported for duty. Within 40 minutes, five operating rooms were up and running.[20]

House physicians throughout the hospital reported to the ECC to assist the nursing and medical staff with triage and patient care. As more ambulance radio reports started to come in, treatment orders for intravenous fluids and medications were given to the various ambulance service units working at the Park. Within minutes the first patient, the cameraman who had collapsed with a myocardial infarction, arrived in the emergency department. His arrival was quickly followed by 34 others. Ninety minutes into the event, the ambulance unit that remained stationed at the Park to serve as incident command called the ECC to report: "We are sending all minor injuries to other hospitals, sending the more serious patients to you." This was relayed to the emergency physician in charge, who called the Grady Incident Command unit back and told them: "Send us all the patients you need to; so far this is business as usual for us, and we have more medical staff, nursing staff, support staff, supplies and equipment than I have ever seen."[21]

Throughout the night, nurses in the ECC provided triage, initiated patient care in the various treatment areas, assisted the physicians with care, maintained crowd control, and made trips to the blood bank, pharmacy, and supply areas all in an orderly fashion. Nurses also assisted FBI and other law enforcement officers who were taking statements and collecting evidence. One of the nurses was overheard talking with a medical student who had come to donate blood. "This is what we do as nurses in the ECC," she said. "No reason to make a fuss, get in a rush, this is what we are trained for and what we do every day."

Of the 35 victims who were treated at Grady, 10 underwent 11 emergency or urgent operations in the first 13.5 hours. There was only one fatality among the victims who were taken to Grady, and that was the cameraman who had suffered the myocardial infarct. Among the other patients taken there, 19 were treated and released and 15 were admitted. All those with life-threatening injuries recovered with excellent or good outcomes. These favorable results were attributed to pre-event disaster preparation as well as to the extensive trauma experience and training of the Grady nursing and medical staff.[22,23]

Later that morning after all the victims had been treated and pressure on the nurses and physicians subsided, Grady administration officials arrived in the ECC to conduct a debriefing and to offer continued counseling support for any interested staff. The meeting did not last long as the nurses agreed that the event had been handled well.

NURSING RESPONSE EFFORTS FOLLOWING THE BOMBING

In a retrospective analysis of the medical response to the bombing victims at Grady, the successful outcomes were attributed to highly effective patient management. The process followed implementation of the disaster plan within the ECC and throughout the hospital. The rapid arrival of the 126 nurses who had been called in the middle of the night, exemplified their dedication, sense of duty, and preparedness for work during a disaster. Grady nurses had been especially prepared to respond in case of an event during the Olympics. Nurses had collaborated with physicians and hospital administrators in the development of the hospital-wide disaster plans, were familiar with phone call fan-out procedures, and had participated in several disaster drills in the weeks preceding the games.[24]

The ability of nursing personnel to rapidly adapt to abrupt changes in practice routines is a major factor in a successful disaster response.[25] Nurses working in emergency settings learn quickly that flexibility and adaptation is a fundamental skill contributing to efficient patient care. Nurses working in other areas of the hospital, where care may be more predictable, may experience greater challenges when required to quickly change the order of operations demanded during a disaster. However, nurses in all parts of the hospital demonstrated the ability to respond flexibly to the emerging needs. Throughout the night they made room for new trauma patients, while still maintaining a high level of care for those already hospitalized.

Another important factor that leads to a successful disaster response is effective communication and collaboration among the healthcare team.[26] Throughout the early morning hours the Grady nursing staff collaborated with law enforcement officials, emergency medical service (EMS) providers as they transferred patients from the scene into the ECC, physician colleagues, and other staff throughout the hospital. Nurses conveyed vital information regarding evolving patient conditions and worked effectively with unfamiliar medical staff who had volunteered to assist to expedite patient care.[27] These efforts also contributed to the efficient, high-quality care provided to the blast victims.

Prior to 2001, nurses working in civilian hospitals had limited opportunities to gain experience or training in the care of blast victims or in the treatment of mass casualties unless they had prior military or Public Health Service Corps experience.[28–30] Although many of the ECC nurses had completed training in advanced trauma nursing through the Emergency Nurse's Association, content on the care of blast victims was not included at that time.[31] Despite this limitation, Grady ECC nurses were able to adapt their knowledge to effectively triage and monitor patient conditions while awaiting transfer to surgery or to intensive care and other medical units.

LESSONS LEARNED

The excellent medical outcomes of the bombing event were abetted by adequate surge capacity staffing and quick adaptations by the nursing personnel. As a result, there were no major problems identified in the nurses' response. However, because of the deluge of telephone calls to Grady Hospital, usual telephone communication methods were impeded.[32] Nurses not prepared to act in emergency conditions expressed concern about not being able to telephone family members to check on their safety and to reassure them of their own. Similarly, some nurses left the hospital worried for their personal safety. These issues highlight the importance of preplanning for disaster events to ensure an adequate nursing workforce and to reduce fears that may hamper operations.

Caring for mass casualty trauma patients in the absence of focused training pointed to the need for continued nurse education in disaster triage and blast injuries. This was later addressed by faculty from the department of surgery, who collaborated with nursing professionals and the American College of Surgeons to develop an advanced nursing trauma life support course, Advanced Trauma Care for Nurses (ATCN) (http://traumanurses.org). The curriculum is similar to the Advanced Trauma Life Support Course (ATLS) that is commonly required for physicians providing emergency or trauma care. The ATCN course is now required for nurses who work in Grady's ECC trauma unit.

OVERVIEW OF THE ROLE OF NURSING DURING DISASTERS

Addressing the medical needs of individuals and families in the context of their environments has long been recognized as essential to a nurse's effectiveness during disaster conditions.[33] In fact, the emergence of nursing as a profession took place in an environment of war and disaster. The founding of the profession is traceable to a group of women led by Florence Nightingale during the Crimean War in 1854. The women had volunteered to tend to sick and wounded British soldiers who had been suffering in filthy and neglected conditions in battle zone camps. Since then nurses have played a vital role in preparedness and response in a variety of disaster situations.[34] It was nurses who through volunteer efforts staffed civilian hospitals throughout Europe and the United States during World War I.[35]

By the start of World War II nurses had begun to formally enter the military to staff mobile army surgical hospital (MASH) units where they applied skills in disaster triage and advanced trauma care.[36] After the war, nurses continued to serve throughout the United States and abroad in the National Public Health Service Corps. As officers in the Corps, nurses became involved in national leadership positions, contributing to the development of public health policies regarding disaster preparedness and response. Other nurses were employed in direct service roles as they deployed with disaster response teams.[37]

With the recognition of emergency medicine as a distinctive field in the 1970s (http://www.aaem.org/aboutaaem/history.php), specialized training in the care of patients with traumatic and emergent medical conditions became available to civilian nurses.[38,39] However, it was not until after 9/11 that formalized nursing preparation and validated clinical competencies for nurses in disaster preparedness and response became available. Competency programs were created by the International Nursing Coalition for Mass Casualty Education in 2003 and the National Organization of Nurse Practitioner Faculties in 2007. Moreover, graduate programs for nurse leaders in disaster preparedness and response were also established during this period.[40]

Currently, nurses can choose from a range of continuing education, interdisciplinary disaster preparedness, and response courses offered at national, state, and local levels. Prime examples include the National Disaster Life Support Courses (NDLS) (http://www.ndlsf.org), National Incident Command System Courses offered through the Federal Emergency Management Agency (FEMA) (http://www.fema.gov/emergency/nims/NIMSTrainingCourses.shtm), and Advanced Hazmat training.[41] These courses prepare nurses to respond to a variety of disaster conditions including natural disasters and those occurring from biological, chemical, radiologic, toxicologic, and explosive acts of terrorism. Such training ensures that nurses understand national emergency response standards for effective functioning during a disaster.[42,43]

Nurses perform many roles in disaster preparation and response including mitigation, surveillance, preparedness planning, and response and recovery efforts. They engage in mitigation efforts by participating in mass immunization programs to reduce the threat of infectious disease or through the dispensing of prophylactic medications. By understanding unusual presentations associated with biological, chemical, or infectious disease exposures nurses assist in surveillance and early detection of health risks both prior to and following disasters. They also play an important role in mitigation through educational and public health initiatives.[44]

A nurse's contribution includes not only hands-on activity but also the ability to provide essential information to others. For example, by understanding the national incident command structure (NIMS) nurses can prepare others to ensure that interagency collaboration runs smoothly, communication failures are prevented, and systems and operations remain functioning.[45] Through continuing education and practice, nurses gain skills that can help prepare others to protect themselves and care for injured and exposed patients during disaster events. Imparting their knowledge to others about disaster response, including mass casualty triage, can thus enhance the overall process.[46]

Once a disaster has occurred, nurses are commonly depended on as the most flexible of all healthcare providers in addressing critical needs.[47] A nurse's clinical, communication, and managerial skills allow him or her to fill a variety of roles to keep systems running. With knowledge of community systems, nurses can provide infrastructure support, collaborate and coordinate interdisciplinary care efforts, increase surge capacity, and assist in resource allocation.[48] Both in hospitals and in the field nurses assist in triage, mass casualty and surge care, trauma care, and early recognition of chemical, biological, or radiologic exposure.[49]

During disasters nurses provide treatment, medication, and patient transport. Advanced preparation allows nurses to adjust practices during mass casualty surges to ensure that standards of care can be maintained until conditions return to normal.[50] By understanding the disaster paradigm—detection, incident command, scene security and safety, assessment of hazard, triage and treatment, evacuation and recovery—nurses can respond effectively.[51]

Nurses may also perform other important functions during a disaster including setting up and staffing shelters for the injured and displaced, administering medicines, distributing supplies, and assisting in the evacuation of vulnerable and medically fragile patients.[52] Finally, nurses provide emotional support to victims and their families, which continues long into the recovery period.[53]

During the recovery phase nurses provide education and counseling, may distribute food and water, and assist individuals and families in accessing needed services. At an institutional and community level, nurses play a role in supporting systems until they are returned to normal operations. Nurses also contribute expertise in evaluating response efforts to identify gaps and develop solutions to prevent future problems.[54]

As communities readjust during the recovery phase nurses provide continued surveillance to identify hazardous conditions or emerging health risks. They may also institute infection control measures to minimize public health risks. In the wake of a disaster nurses provide continued patient care within hospitals, provide shelters and other community settings, and may also coordinate relief efforts through interagency communication.

Thus, nurses interact with the full range of other responders and interested parties, from law enforcement officers, firefighters, the military, and Hazmat teams to public health officials, EMTs, paramedics, physicians, other healthcare professionals, patients, and their families. This demands cooperation, collaboration, effective communication, and a willingness to adhere to appropriate chains of command and national standards for disaster response. Maintaining consistent standards may require functionally equivalent care when conventional care is not possible, a goal during disaster conditions that has been underscored by the Institute of Medicine.[55] Nurses are well positioned to achieve this goal in view of the breadth of their responsibilities and their experience in interdisciplinary response training.

CONCLUSION

Nurses are equipped with an unusually broad set of skills, attributes, and responsibilities. As the largest group of healthcare professionals in the United States, they provide critical manpower in a large variety of emergency situations.[56] But coordinated and effective efforts can only be a product of proper preparation. Besides knowledge about how to act in emergencies, preparation includes planning in advance for a nurse's own safety and the safety and care of family members, which also can forestall unwillingness to report to duty.[57] Preparedness incorporates as well the attribute of versatility, which better enables the nurse to adapt to disaster care conditions.[58] These capabilities contribute to the extraordinarily high esteem accorded to the nursing field: Annual Gallup surveys consistently rank nurses as more honest and ethical than members of any other profession.[59]

Nurses who graduated from educational programs prior to 2003 (before disaster competency and educational standards were established for nursing curricula) were less well prepared unless they had voluntarily completed disaster training through continuing education.[60,61] Many states require nurses to complete annual continuing education courses to maintain their licensure. Still, the requirements may not include training in advanced trauma or disaster preparation and response unless the nurse is employed in an emergency or public health capacity. Therefore, manpower gaps in disaster preparation continue to exist within the profession.[62] These gaps could be reduced if every registered nurse attended a basic disaster response course.

The effective care that occurred following the Olympic Park bombing derives in part from Grady's being a level 1 trauma center and its nursing and medical staff being well versed in trauma care.[63] Had the bombing occurred in a rural setting or near a hospital that lacked in-house trauma services, patient outcomes might have been dramatically different. The aftermath of Hurricane Katrina demonstrated that systems could easily become overwhelmed, even in trauma centers, if manpower or adequate preparation is lacking.[64] It is not a question of "if" but "when" the next disaster will occur in the United States. Efforts to respond and recover will match the needs only if all nurses and other relevant medical personnel are prepared.

A dismaying postscript about disaster readiness at Grady and elsewhere appeared a decade after the Olympic Park bombing. In 2006, Arthur Kellermann recalled the 1996 event and the efficiency of the medical response in which he participated as an emergency physician. Subsequently he became Chairman of the Department of Emergency Medicine at Emory University and had come to doubt that the response could now be replicated:

"America's emergency and trauma care system has deteriorated to an alarming degree," he wrote. "Today, Grady's trauma service is overflowing, its operating rooms are always full, and its emergency department is packed with patients. Under such circumstances, it is hard to see how the hospital could manage another bombing on the scale of the Olympic Park event, much less attacks of the magnitude observed in Madrid, London, and Mumbai."[65]

Kellermann concluded with a plea for more financial and other support for training, continuing education, and credentialing of emergency care professionals. Key among them, of course, would be nursing professionals.

NOTES

1. Booth W, Heath T. Deadly blast rocks Atlanta Plaza; explosive device injures scores in Olympic Park. Washington Post, July 27, 1996.

2. Adler J, Smith VE. The dream turns to nightmare [cover story]. Newsweek, 1996;128(6):26.

3. Ibid.

4. Duffy B, Goode E. Terror at the Olympics [cover story]. US News & World Report 1996; 121(5):24.

5. Anderson GV Jr, Feliciano DV. The Centennial Olympic Park bombing: Grady's response. J Med Assoc Ga 1997;86(1):42–46.

6. Feliciano DV, Anderson GV Jr, Rozycki GS, et al. Management of casualties from the bombing at the Centennial Olympics. Am J Surg 1998;176(6):538–543.

7. Anderson.

8. Cantwell JD. The Olympic medical experience: An overview. J Med Assoc Ga 1997;86:13–14.

9. Sharp TW, Brennan RJ, Keim M, et al. Medical preparedness for a terrorist incident involving chemical and biological agents during the 1996 Atlanta Olympic Games. Ann Emerg Med 1998;32(2):214–223.

10. Duffy.

11. Starr M, Smith VE. Is Atlanta ready yet? Newsweek, July 17, 1995.

12. Sharp.

13. Adler.

14. Starr.

15. Duffy.

16. Sharp.

17. Feliciano.

18. Anderson.

19. Ibid.

20. Ibid.

21. Feliciano.

22. Ibid.

23. Anderson.

24. Ibid.

25. Slepski LA. Emergency preparedness and professional competency among health care providers during hurricanes Katrina and Rita: Pilot study results. Disaster Manag Response 2007; 5(4):99–110.

26. Gebbie KM, Peterson CA, Subbarao I, White KM. Adapting standards of care under extreme conditions. Disaster Med Public Health Prep 2009;3(2):111–116.

27. Anderson.

28. Gebbie K, Qureshi K. A historical challenge: Nurses and emergencies. Online J Issues Nurs 2006;11(3):2.

29. Stein M, Hirshberg A. Medical consequences of terrorism. The conventional weapon threat. Surg Clin North Am 1999;79(6):1537–1552.

30. Debisette AT, Martinelli AM, Couig MP, Braun M. US Public Health Service Commissioned Corps Nurses: Responding in times of national need. Nurs Clin North Am 2010;45(2): 123–135.

31. Broering B., editor. *TNCC: Trauma Nursing Core Course*. 6th ed. Des Plaines, IL: Emergency Nurses Association; 2007.

32. Anderson.

33. Garfield R, Dresden E, Rafferty A-M. Commentary: The evolving role of nurses in terrorism and war. Am J Infect Control 2003;31(3):163–167.

34. Gebbie, Qureshi.

35. Garfield.

36. Gebbie, Qureshi.

37. Debisette.

38. Gebbie, Qureshi.

39. Broering.

40. Littleton-Kearney MT, Slepski LA. Directions for disaster nursing education in the United States. Crit Care Nurs Clin North Am 2008;20(1):103.

41. Walter F, editor. *Advanced Hazmat Life Support*. 3rd ed. Tucson, AZ: University of Arizona Press; 2003.

42. Gebbie, Qureshi.

43. American Nurses Association. Who will be there? Ethics, the law, and a nurse's duty to respond, 2010. Available at http://www.nursingworld.org/MainMenuCategories/WorkplaceSafety/DPR/Disaster-Preparedness.pdf, accessed June 17, 2012.

44. Gebbie, Qureshi.

45. Ibid.

46. Dallas C, Coule P, James J, et al., editors. *Basic Disaster Life Support Manual*. Chicago, IL: American Medical Association; 2004.

47. Gebbie, Qureshi.

48. Debisette.

49. Gebbie, Qureshi.

50. Garfield.

51. Dallas.

52. Debisette.

53. Littleton-Kearney.

54. Gebbie, Qureshi.

55. Altevogt B, Stroud C, Hanson S, et al. *Guidance for Establishing Crisis Standards of Care for Use in Disaster Situations: A Letter Report*. Washington, DC: Institute of Medicine; 2009.

56. Littleton-Kearney.

57. Gebbie, Qureshi.

58. Slepski.

59. American Nurses Association. Public ranks nurses as most trusted profession, 2010. Available at http://www.nursingworld.org/FunctionalMenuCategories/MediaResources/PressReleases/2010-PR/Nurses-Most-Trusted.aspx, accessed, June 17, 2012.

60. Littleton-Kearney.

61. Nursing Emergency Preparedness Education Coalition. Educational competencies for registered nurses responding to mass casualty incidents, 2003. Available at http://www.nursing.vanderbilt.edu/incmce/competencies.html, accessed June 17, 2012.

62. Littleton-Kearney.

63. Anderson.

64. Slepski.

65. Kellermann AE. Crisis in the emergency department. N Engl J Med 2006; 355: 1300–1303.

6

THE ROLE OF THE DENTIST

David L. Glotzer

THE BEIRUT BARRACKS ATTACKS

In 1983 a contingent of US Marines was sent to Lebanon as part of an international peacekeeping force to help stabilize the country, which had been torn by a civil war between Christians, their ally Israel, and the Muslim population. It was not easy for this force of US and European troops to keep the peace in a nation of fractious religious and political loyalties. The hope was that if a fragile peace could hold in Beirut, the nation's 8-year civil war might end. Broad acceptance of a national government could then be possible.

The 1st Battalion 8th Marines, under the US 2nd Marine Division, consisting of 1200 men, set up its local headquarters at the Beirut International Airport, operating under their given limited rules of engagement. On October 23, 1983, about 6:20 A.M., a yellow Mercedes-Benz truck drove toward the airport. The truck had been substituted for an expected water delivery vehicle that had been hijacked. The truck turned onto an access road leading to the Marines' compound and circled a parking lot. The driver then accelerated and crashed through a barbed wire fence around the parking lot, passed between two sentry posts, rammed through a gate, and barreled into the lobby of the Marine headquarters. By the time the two sentries had locked and loaded their weapons, the truck was already inside the building's entryway.

The suicide bomber detonated his explosives, which were equivalent to 12,000 pounds of TNT. The force of the explosion collapsed the four-story cinderblock building into rubble, causing crushing injuries to many inside. The blast sheared the bases of the concrete support columns and the airborne building collapsed. A massive shock wave and ball of fire followed.

About 20 seconds later, an identical attack occurred against the barracks of the French 3rd Company of the 6th French Parachute Infantry Regiment. Another suicide bomber had driven his truck down a ramp into the building's underground parking garage and detonated his bomb, leveling that headquarters.

Local Planning for Terror and Disaster: From Bioterrorism to Earthquakes, First Edition.
Edited by Leonard A. Cole and Nancy D. Connell.
© 2012 John Wiley & Sons, Inc. Published 2012 by John Wiley & Sons, Inc.

In the attack on the American barracks, the death toll was 241 American servicemen: 220 Marines, 18 Navy personnel, and 3 Army soldiers. Sixty Americans were injured. In the attack on the French barracks, 58 paratroopers were killed and 15 injured. In addition, the elderly Lebanese custodian of the Marines' building was killed in the first blast. The wife and four children of a Lebanese janitor at the French building also were killed.

This was the deadliest single-day death toll for the US Marine Corps since the Battle of Iwo Jima in World War II and the deadliest single-day death toll for the US military since the 243 killed on January 31, 1968, the first day of the Tet offensive in the Vietnam War.[1]

The only on-scene physician, a naval officer, was instantly killed and buried in the rubble, along with 18 Navy hospital corpsmen. As it happened, the two Navy dental officers assigned to the 24th Marine Amphibious Unit were unhurt. They ran to the site and subsequently coordinated emergency trauma care with 15 surviving hospital corpsmen. They treated 65 casualties in the first 2 hours following the explosion. Lieutenants Gilbert Bigelow and James J. Ware would later be awarded Bronze Stars for their leadership and the emergency medical services they provided. Additional dental personnel aboard the USS Iwo Jima stationed offshore eventually joined the medical teams sent to provide care and support for the airport blast survivors.[2]

Drs. Ware and Bigelow had no idea of the carnage they would have to deal with as they rushed to the attack scene. Prioritizing on the run, they quickly found themselves in charge and decided that "one would go down to the blast site, and the other one would triage in the area," said Dr. Bigelow. "We figured we would go into triage and do the best we could with what we had, sorting bodies and individuals out and either getting them to the helicopter zone and directly out or up to Jim," he continued. "When you look around and you're the one standing, you're the one standing."

"We had physicians who came from the ship but Gil and I orchestrated what was going on," said Dr. Ware. They worked through the long and bloody day of the Beirut barracks bombing, triaging patients, and handling the morgue. "My question was, does this patient stay here? Does he need to go now? This is the hard part," said Ware. Do I have the medical qualifications to make that call? You begin to make the decision. I did what I had to do."

Colonel Timonthy J. Geraghty the Marine commander in Beirut wrote in his detailed narrative of the events of that day, "Dr. Gil Bigelow had hooked up with his assistant, Dr. Jim Ware and some corpsmen and set up two separate aid stations. Jim was placed in charge while Dr. Bigelow took four corpsmen, rushed back to the BTL (battalion landing) site and set up the triage station. Scores of wounded would eventually cycle through these stations and many owe their lives to these dedicated professionals."[3]

"He was Butch Cassidy," Ware said of Bigelow in a telephone interview some 25 years after that day of tragedy. " 'I'm the Sundance Kid. Like a good big brother' " he said, 'we're going to have to take the bull by the horns.' I give John Hudson, the physician, the friend who was killed, credit for the planning, for making sure we were all integrated into the medical plan. I give Gil credit for his maturity."[2]

DENTISTS AS KEY RESPONDERS

Whenever there are more casualties than the medical care system can accommodate, some system of triage (sorting) must be established. What was clearly demonstrated in this tragedy was that because of their training, these dentists were able to lead and augment personnel in providing triage and treatment. Simply stated, appropriately trained dentists can fulfill these additional functions.

The bombing of the marine barracks was a "sudden-impact" mass casualty event (MCE). This category may include not only large explosions but also airplane or train crashes and natural disasters such as earthquakes or tornados. But all sudden-impact MCEs require immediate responses in the form of triage and life-saving interventions. In addition, while destroying local medical facilities and medical personnel, they require large-scale movement of the survivors to alternate sites.

In contrast, exposure to bioterrorist agents such as anthrax or smallpox microorganisms, or a pandemic agent such as the H1N1 influenza virus, may be described as a "developing-impact" MCE. The number of people affected by the event could gradually increase exponentially. The number could decline due to treatment and prophylactic efforts, but then increase again as a result of additional exposures to the disease agent.[4]

We know that during civilian disaster events such as Hurricane Katrina, people may be without power, shelter, communication, food, and water. Emergency response capabilities may be overwhelmed and local emergency medical services may be unable to gain access to victims. Healthcare facilities might be damaged during the impact, which would impair their ability to provide care. The needs of the situation may be massive and overwhelm the resources available.

A situation becomes a medical disaster when the ability to provide healthcare is overwhelmed. Thus, preparedness should assume an *all-hazards* approach rather than focus on a particular type of incident. It may then fall to nontraditional healthcare workers to provide services ordinarily performed by physicians. In a word, disasters can inflict damage beyond the conventional resources available to a community.[5]

Unfortunately, dentists are rarely called upon explicitly to provide care during a disaster event. In fact, they are well suited to assist with a variety of emergency functions when normal delivery of healthcare is strained or disrupted. They possess basic healthcare skills together with experience managing apprehensive patients that could be of great value during disaster conditions.

Dentists are not generally incorporated in the planning phase for such events. Until they are, their skills and understanding of the local community are less likely to be employed during an actual event.

In conducting their practices, they routinely engage in quick critical thinking, innovation, and problem solving. These capabilities could be very useful during a disaster when confronting, for example, a shortage of supplies or confusion in staff communication. Although no two disasters are exactly the same, dental professionals often must be able to improvise and adapt to stressful situations.[6,7]

Functions and Services during an Event

Functions and services that a dentist could perform, albeit with some additional training, include

- *Medical Care Augmentation.* This might involve treatment of cranial and facial injuries, providing or assisting in administration of an anesthetic, starting intravenous lines, performing appropriate surgery and suturing, assisting in shock management, taking medical histories, and providing cardiopulmonary resuscitation.
- *Surveillance, Notification, Monitoring.* Since there is an incubation period before the clinical manifestations of a disease that may have resulted from a bioterrorist attack become apparent, and because dental offices are distributed across the community,

dentists can serve as an excellent surveillance resource. They can detect characteristic intraoral or cutaneous lesions and report them to public health authorities. They also may be able to detect unusual patterns of employee absences or patient cancellations, such as those due to flu-like symptoms at an unseasonal time of year. Salivary and/or nasal swabs may yield important diagnostic or treatment information and can be collected by dentists for laboratory testing to determine diagnoses when necessary or to monitor treatment progress.

- *Immunizations and Medications.* When rapid inoculation or vaccination of the public is required, dentists may be recruited to assist in a mass inoculation program. Physicians and nurses alone might be unable to fulfill the requirements of such a program quickly enough for successful prevention of the spread of disease. If the population requires mass treatment, preventive medication, or both, pharmacies' capabilities may become overwhelmed quickly. Dentists could be called on to prescribe and dispense chemotherapeutic medications for the public, and they can monitor patients for adverse reactions and side effects.

- *Triage.* Whenever there are a greater number of casualties than the medical care system can accommodate, or whenever medical care resources are overwhelmed, priorities for treatment must be established. Appropriately trained dentists can fulfill this role. This has especially been shown to be true in the military.

- *Dental Offices.* These are located throughout any given community and have many of the resources that are found in hospitals: sterilization equipment, air and gas lines, suction equipment, radiology capabilities, instruments, and syringes. They may be called on to serve as local "mini-hospitals" when local hospital facilities become overwhelmed or when the concentration of patients is to be avoided, as in attacks involving contagious agents. Concentrations of patients and healthcare providers also may present tempting secondary targets to attackers. In some scenarios, decentralization of medical care may be the most appropriate response.

- *Decontamination and Infection Control.* Dentists and dental auxiliaries are well versed in infection control procedures. They can apply their knowledge in reducing the spread of infections—between patients and between patients and caregivers in mass disasters. The decontamination of casualties, when appropriate, can be accomplished effectively by dental personnel. Dentists who have worked in a hospital setting are especially equipped to provide services that require a close working relationship with physicians. Consequently, in the event of a bioterrorist attack, dentists may be called on to fulfill several functions: education, risk communication, diagnosis, surveillance and notification, treatment, distribution of medications, decontamination, and sample collection.

- *Forensics.* Dentistry plays a significant role in this process. By identifying the victims of crime and disaster through dental records, dentists assist those involved in crime investigation. They are always part of a bigger team and are dedicated to the common principles of all those involved in forensic casework: the rights of the dead and those who survive them. The most common role of the forensic dentist is the identification of deceased individuals. The bodies of victims of violent crimes, fires, motor vehicle accidents, and workplace accidents can be so disfigured that identification by a family member is neither reliable nor desirable. Dental identifications have always played a key role in natural and manmade disaster situations and particularly in MCEs. Therefore, dentists trained in forensic odontology may very well work closely with local Disaster Mortuary Operational Response Teams, known as DMORTs.

- *Dental Auxiliaries.* Dental office auxiliaries can provide important assistance in the initial response to a major bioterrorism attack. All are familiar with and can serve in administrative functions, managing medical records and handling patient flow. With additional training, dental hygienists can provide valuable assistance in mass disaster responses by providing clinical services beyond those they ordinarily supply, possibly including administration of inoculations and vaccinations. Dental assistants can continue to aid dentists, even in tasks different from those they usually perform. Clerical staff can provide an important link in communications between dentists and other clinicians.[6–8]

Matching a Sense of Responsibility with Training

Many dentists feel strongly that they have an ethical responsibility to respond in a disaster situation and that other clinicians would be receptive to their assistance. They feel that they have clinical skills that could be useful in a catastrophic response effort. They are agreeable to additional training. (They do not necessarily believe, however, that this training should be required as a condition for licensure, board certification, or credentialing.) Barriers to training include concerns about risk and malpractice, the cost of training, the time involved in training, and the cost for the time in training (e.g., lost revenue and continuing medical education time). Dentists are not concerned about whether they can learn and retain these skills. We can conclude that improving healthcare preparedness to respond to a terrorist or natural disaster requires increased efforts at organization, education, and training.

Dentists are willing to increase their knowledge base if it is possible to create a mutually positive environment, to minimize cost and disruption while maximizing preparedness. To most efficiently and effectively use homeland security dollars, a dialogue must begin between the dental profession, professional dental societies, and the US Department of Health and Human Services to determine the best training strategies.[10,11]

Dentistry's role in disaster response is relatively new and still evolving. After the events of September 11, 2001, professional dental organizations, schools of dentistry, individual authors, and other entities have emphasized dentistry's role in emergency preparedness and response, bio-response, vaccination/immunization, disease surveillance, and other related activities. The Academy of General Dentistry (AGD) published an AGD Impact issue that highlighted roles played by AGD dentists in the September 11 attacks. In June 2002, the American Dental Association (ADA) convened a workshop titled "Dentistry's Response to Bioterrorism" where it developed a consensus report outlining the various roles dentistry might play in response to public health emergencies.[7]

In March 2003, the ADA and the US Public Health Service (USPHS) sponsored a conference on Dentistry's Role in Responding to Bioterrorism and Other Catastrophic Events. The conference was attended by oral health and dental professionals from the Department of Defense, USPHS, and numerous other federal agencies and divisions, and representatives from organized dentistry, dental academia, and state, local, and other public health entities. At that event, US Surgeon General, Vice Admiral Richard Carmona emphasized, "Dentists have a role in emergency response because they have the patient care skills, medical knowledge and communication skills."[12]

He opened the 2-day conference with an overview of the then current threat and invited the dental profession to participate in the war on terror. "I will consider dentistry an equal partner with all the other health professions at the table as we approach these challenges before us," he said. "I will probably ask for your assistance and guidance. Let us know how you think you can best serve your country." More than 300 registrants, almost all of them dentists, welcomed the Surgeon General's dialogue aimed at shaping a strategy for professional service in the war on terrorism. "Who would have ever thought there'd be such a huge demand for forensic dentistry?" he said referring to the terrorist attacks of September 11, 2001, and the lead role of forensic dentists in identification. "It's really terrible but that's the world we live in. Who would have ever thought we'd be dealing with germs as weapons, planes as weapons? What if we have to immunize a lot of people because of an impending threat," said Surgeon General Carmona. "Could not dentists be involved? Could not dentists be involved in taking histories, screening patients, dealing with out-patient issues? Absolutely. I think we're only on the very brink of how much dentistry can be involved as a response force as part of our partnership and team as we prepare our country for these inevitable threats."[12]

In April 2006, the *Journal of the American Dental Association* published an article that spelled out the process that occurred in Illinois, where an expanded role of the "Dental Emergency Responder" had been written into the Illinois State Dental Practice Act. The article also described a plan developed by the American Medical Association (AMA) through its Center for Public Health Preparedness and Disaster Response, called the National Disaster Life Support Education Consortium (NDLSEC) program. The mission of the NDLSEC is to promote excellence in education and training in disaster medicine for all health professionals, and thereby create a network to respond to a public health emergency. Within the NDLSEC protocols for all-hazards disaster response and their competencies for health professionals in disaster medicine and public health preparedness, there are clear roles that can be filled by properly trained dentists. These include, among others, being able to conduct hazard vulnerability assessments in a community, identifying psychological reactions exhibited by victims, understanding signs and symptoms of disease and injury likely to be associated with disasters, demonstrating the ability to apply and adapt clinical knowledge and skills in management of victims, and monitoring the safety and public health concerns of the environment, and more.[13]

In 2007, the ADA's House of Delegates passed a resolution to help guide the association's position on emergency preparedness and disaster/response issues. The resolution was the result of a May 2007 mega-issue discussion by the ADA's Council on Dental Practice. It called on the Association to

- help dentists become more effective responders to natural disasters and other catastrophic events
- provide leadership in national, state, and community disaster planning and response by increasing efforts that put disaster preparedness into practice
- promote multidisciplinary disaster courses and education that train dentists and dental staff members in the handling of declared emergencies
- advocate for national emergency-preparedness solutions through research, public policy, and legislation[14]

With dentistry's role defined and Association policy firmly established, the ADA has moved forward with a series of steps to promote dentistry's role in responding to disasters and other public health emergencies. These steps incorporate consideration of the following challenges:

- Credentialing systems affecting dentists, dental assistants, dental hygienists, and other health professionals interested in supporting the medical and public health response to a major disaster
- Education and training to help dental volunteers establish and plan for their roles in caring for large numbers of patients during public health emergencies
- Civil liability protection and licensure accommodation for in-state dentists and those traveling across state lines for the good-faith treatment of casualties during public health emergencies[15]

To ensure dentistry's place in federal disaster response, the ADA also has been urging lawmakers to elevate dentistry's role in the federal disaster response framework. In early 2011, Representative Michael Burgess [R-TX] introduced H.R. 570, the Dental Emergency Responder Act of 2011. This legislation would raise dentistry's role in federal disaster planning by definition and make clear that dental schools are eligible to apply for public-health and medical-response training grants. It also would strengthen America's medical surge capacity by taking advantage of the extensive education, training, and professionalism of dentists. The act requires no new federal money, poses no new restrictions on funds now being spent, and places no new mandates on dentists. After passage in the House in March, the bill was sent to the Senate for consideration.[16]

DEVELOPING THE DENTAL SCHOOL CURRICULUM

Another facet of the discussion of the role of the dentist in preparedness and disaster is the responsibility of the nation's dental schools. The pre-doctoral dental curriculum should include the subject of bioterrorism, preparedness, and a possible medical response. New dentists should have the basic information required to function effectively in an MCE, as well as quick-reference materials to use when the need arises. Continuing education programs for established dentists should also be developed.

In early 2002, the leadership of the New York University College of Dentistry (NYUCD) sought to define a role for the dental profession in response to the new threats the United States was facing. Proposals included expanding the dental school curriculum to include bioterrorism studies and training for response to a public health disaster.[17] At the same time the ADA began to focus on defining the role of dentists in responding to a bioterrorism attack. The Association's specific areas of concern were (1) what preparation was required to respond to an attack, including the appropriate role of dental schools in training dentists as responders, and (2) to what extent, if any, training should be mandated.

The outcome of a consensus meeting was that an understanding of bioterrorism should be part of the pre-doctoral curriculum. This would include a basic level of training to provide treatment and preventive measures under the direction of a responsible emergency response agency. The ADA and the American Dental Education Association (ADEA) then cosponsored a workshop entitled "Terrorism and Mass Casualty Curriculum Development." Participants recommended that core competencies be taught to all dental students to familiarize them with biological and chemical agents that might be used in an attack, prepare them to respond to an event, and create a cadre of healthcare professionals who could serve as a source of surveillance information in the event of an attack.[18]

In addition to providing education for the dental profession, dental schools can serve as assets during the actual response to an attack. Since most schools are associated with academic health centers, whose hospitals may be overwhelmed with demands for care, they may be mobilized as hospital annexes. Dental students and faculty may serve in the dental school itself, or in the hospital, or in the field as primary responders. Dental schools also can be repositories for pre-stocked supplies and equipment.[8]

Another potential asset is the local dental society. It could develop plans for the dental response to a terrorist attack or natural disaster that can be integrated into each community's mass disaster response plan. Educational programs for dentists should be developed to prepare them to provide services in an emergency. Ability to serve in community responses would be greatly enhanced if dental professionals have already received training in their pre-doctoral education.

There are predictable challenges when developing a curriculum for pre-doctoral students in our dental schools. Among the facts and assumptions that need to be addressed are (1) catastrophic events take different forms, none of which is exactly predictable. (2) Most dentists are less likely to be first responders and more likely to be community resources for preparation and an integrated response. (3) Dentists, by virtue of their overall education, have skills that can be applied directly or adapted easily when needed. (4) Dentists who are willing to be first responders would need additional education beyond their pre-doctoral education. (5) Dentists, as members of their community, would be accountable for developing personal protection plans to ensure their safety and survival so that they would remain a community resource. (6) Dentists, regardless of their education, face the ethical dilemma of deciding whether to join in their community's response or stand aside. Based on these assumptions, NYUCD developed the following competencies for dental graduates:

> Competency 1: Describe the potential role of dentists in the first/early response in a range of catastrophic events.
>
> Competency 2: Describe the chain of command in the national, state, and/or local response to a catastrophic event.
>
> Competency 3: Demonstrate the likely role of a dentist in an emergency response and participate in a simulation/drill.
>
> Competency 4: Demonstrate the possible role of a dentist in all communications at the level of a response team, the media, to the general public, and to patient and family.
>
> Competency 5: Identify personal limits as a potential responder and sources that are available for referral.
>
> Competency 6: Apply problem solving and flexible thinking to unusual challenges within the dentist's functional ability and evaluate the effectiveness of the actions that are taken.
>
> Competency 7: Recognize deviations from the norm, such as unusual cancellation patterns, symptoms of seasonal illnesses that occur out the normal season, and employee absences, that may indicate an emergency.[17]

FITTING INTO THE CONSTELLATION OF HEALTHCARE PROFESSIONALS

In anticipation of further terrorist and disaster events, it is an absolute duty of responsible agencies in all healthcare related professions to help enhance the medical response system.

To be properly prepared for such events, trained, willing, and available personnel should be identified *beforehand* to insure optimal survival rates and healthcare outcomes. To meet such medical and public health manpower needs, a surge capacity must be developed from relevant professionals who could complement the traditional medical and public health workforce. Clearly, these additional resources require familiarity with health/public health principles and practice.

However, mobilizing "surge" or supplemental healthcare professionals within an organized pre-planned system is challenging. Hospital administrators involved in responding to the World Trade Center tragedy reported that they were unable in some cases to verify the volunteer's credentials, training, skills, and competencies. This underscores the need for pre-event organization of a qualified health workforce that could provide supplemental value.[19]

At least two systems are now in place and are available to emergency healthcare planners to insure that surge volunteers have a professional license, have recognized professional skills, and may have additional preparedness training. Congress has recognized the need to make optimum use of volunteer health personnel in an emergency. It authorized the development of an Emergency System for Advance Registration of Volunteer Health Professionals (ESAR-VHP) in the Public Health Security and Bioterrorism Preparedness and Response Act of 2002. The Health Resources and Services Administration (HRSA) was made responsible for assisting each state in establishing a standardized, volunteer registration system. The standardization of state systems gives each state the ability to quickly identify and better utilize health professional volunteers in emergencies and disasters.

The program goal of ESAR-VHP is to have in place state-based systems that together form a national system for efficient utilization of health professional volunteers in emergencies. Verifiable, up-to-date information regarding a volunteer's identity and credentials can allow quick exchanges of health professionals among the states.[20]

The second system that organizes health professionals in anticipation of a mass casualty incident is the Medical Reserve Corps (MRC). It is a specialized component of the Citizen Corps Council (CCC) that offers an opportunity for all healthcare providers to become an integrated part of an emergency response plan. The CCC is a national network of volunteers dedicated to making sure that their families, homes, and communities are safe from terrorism, crime, and disasters. The MRC Program was launched in July 2002 and is overseen by the US Office of the Surgeon General. Its mission is to organize teams of local volunteer medical and public health professionals who can supplement existing community emergency response systems during large-scale events. There are approximately 600 MRC units nationwide.[21,22]

CONCLUSION

The following proposals bear consideration by the dental, medical, and public health leadership:

- Dental curricula should incorporate catastrophic event training in pre-doctoral and graduate programs in all dental schools.
- Dental educators in collaboration with their medical counterparts should develop continuing education programs on catastrophic events, specifically for the practicing dental community.

- Catastrophic event training for dentists should be flexible, and be particularly attentive to the local community medical and public health infrastructure.
- Various other professional leadership groups should be included along with dentists in catastrophic event response training, for example, State Public Health Officer, Office of Emergency Management.
- Assessments should be conducted among individual members of the dental community to determine the best ways to advance their interest in preparing for catastrophic events.

Essentials of disaster response include three changes from a dentist's normal approach to clinical care:

1. A disaster response could include triage, with the aim of doing the greatest good for the greatest number.
2. The public health focus is no longer just treating the individual but enhancing the wellbeing of the larger population.
3. Overall direction of the response may require modification of existing legal, regulatory, and ethical principles and guidelines.[23]

The federal government since September 11, 2001, the anthrax attacks that followed, Hurricane Katrina, and the more recent H1N1 threat has increased the support, development, and national attention given to public health emergency preparedness. Bioterrorism, natural disasters, and emerging infectious diseases all have the potential to overwhelm the existing medical system. It is logical that a dedicated and trained supplemental workforce should be developed.

Whatever the challenges to the use of this "surge" or supplemental healthcare resource, such as balancing the legal and ethical concerns, and deciding what is an acceptable level of medical care in a disaster, the dental profession is well suited to help in this role. With proper additional training and integration into the organized healthcare system, dentists can be an important source of this much-needed manpower.

NOTES

1. 1983 Beirut barracks bombing, Wikipedia. Available at http://en.wikipedia.org/wiki/1983_Beirut_barracks_bombing, accessed June 18, 2012.
2. Palmer C. Drs. Ware, Bigelow awarded Bronze Star after Beirut Bombing. ADA News, September 14, 2009.
3. Geraghty TJ. *Peacekeepers at War: Beirut 1983—The Marine Commander Tells His Story*. New York: Putnam Books; 2009:99.
4. Phillips SJ, Knebel A, Johnson K, editors. *Mass Medical Care with Scarce Resources: The Essentials*. Rockville, MD: AHRQ; 2009. Publication No. 09-0016.
5. Sundes K, Birnbaum M. Healthcare disaster management: Guidelines for evaluation and research in the Utstein style. Prehosp Disaster Med 2003;17(3):1–177.
6. Psoter WJ, Alfano MC, Rekow ED. Meeting a disaster's medical surge demand: Can dentists help? J Calif Dent Assoc 2004;32(8):694–700.
7. Glotzer DL, Psoter WJ. Disasters and the surge environment. J Emerg Manag 2006;4(3):47–52.
8. Guay AH. Dentistry's response to bioterrorism, a report of a consensus workshop. J Am Dent Assoc 2002;133:1181–1187.

9. Psoter W, Triola M, Morse D, Rekow E. Enhancing medical and public health capabilities during times of crisis. N Y State Dent J 2003;69(5):25–27.

10. Steigbigel NH, Blaser MJ, Brewer K, et al. Enhancing medical and public health capabilities during times of crisis: A grant. Department of Justice, Office of Justice Programs, 202-DT-CX-K002. New York: 2005.

11. Psoter W, Glotzer D. *A Summary Report on the Expansion of the Role of Dentists and their Enhancement of the Medical Surge Response.* New York: NYU College of Dentistry; 2005.

12. Dentists are equal partners in war on terrorism: Surgeon General. ScienceBlog, posted March 31, 2003. Available at http://scienceblog.com/1353/dentists-are-equal-partners -in-war-on-terrorism-surgeon-general/, accessed June 18, 2012.

13. Colvard MD, Lampiris LN, Cordell GA, et al. The dental emergency responder: Expanding the scope of dental practice. J Am Dent Assoc 2006;137(4):468–473.

14. Furlong A. ADA House defines dentistry's role in disaster response. ADA News, November 20, 2007.

15. Flores S, Mills SE, Shackelford L. Dentistry and bioterrorism. Dent Clin North Am 2003;47(4):733–744.

16. Furlong A. Congress advances dental responder legislation. ADA News, March 15, 2011.

17. More FG, Phelan J, Boylan R, et al. Predoctoral curriculum for catastrophe preparedness. J Dent Educ 2004;68(8):851–858.

18. Chmar JE, Ranney RR, Guay AH, et al. Incorporating bioterrorism training into dental education: Report of ADA-ADEA workshop. J Dent Educ 2004;68(11):1196–1199.

19. Glotzer DL, More FG, Phelan J, et al. Introducing a senior course on catastrophe preparedness into the dental school curriculum. J Dent Educ 2006;70(3):225–230.

20. Health Resources and Service Administrations. Emergency system for advance registra-tion of volunteer health professionals. Available at http://www.publichealthlaw.net/Research/ PDF/ESAR%20VHP%20Report.pdf, accessed June 18, 2012.

21. Medical Reserve Corps. Office of the Surgeon General. Available at http://www.medicalre servecorps.gov/, accessed June 18, 2012.

22. Glotzer DL, Rinchiuso A, Rekow D, et al. The Medical Reserve Corps, an opportunity for dentists to serve. N Y State Dent J 2006; 72: 1.

23. Auf der Heide E. The importance of evidence-based disaster planning. Ann Emerg Med 2006;47(1):34–40.

7

THE ROLE OF THE EMERGENCY MEDICAL TECHNICIAN

Brendan McCluskey and Henry P. Cortacans

THE DC METRO CRASH

On a clear evening in June 2009, Washington DC Metro Train No. 112 left the Takoma station at 4:57 P.M. Five minutes later it struck the No. 214, which had stopped on the same track awaiting the departure of another train from the Fort Totten station. The impact forced the moving train into, then up and over, the last car of the No. 214. The collision compressed the lead car of the No. 112 reducing its interior space by 84%. People were crushed against their seats. Eight passengers were killed along with the train operator, and 90 others were injured.

The first 9-1-1 call was logged at 5:03 P.M. from a train passenger, followed rapidly by calls from other passengers and a passerby. The initial call was forwarded to the Metro Transit Police Department, and at 5:04 P.M., the city Fire Department dispatched three engine companies, two truck companies, two battalion chiefs, the special operations battalion chief, a rescue squad, an ambulance, a medic unit, an emergency medical services (EMS) unit, and a safety officer to the Takoma station. A second series of response units were directed to the Fort Totten station. Two groups of responders then walked along the track toward the crash; rescuers first entered the lead car of the moving train at 5:20 P.M. Ninety passengers were triaged and treated at the site, all within the first 90 minutes after the crash.

The emergency response had been swift and organized. A command post was established nearby on New Hampshire Avenue. The assistant chief of operations for the District of Columbia Fire and Emergency Medical Services Department served as the incident commander. The deputy chief for special operations was in charge of rescue and extrication operations. A battalion chief was in charge of the evacuation of patients from the trains. Another battalion chief led medical operations, and the DC Police and the Metro Police maintained a liaison at the command post.[1]

Local Planning for Terror and Disaster: From Bioterrorism to Earthquakes, First Edition.
Edited by Leonard A. Cole and Nancy D. Connell.
© 2012 John Wiley & Sons, Inc. Published 2012 by John Wiley & Sons, Inc.

Emergency personnel—numbering more than 200 within 2 hours of the crash—worked through the night. They used rescue equipment to free trapped passengers and search for bodies. A crane was needed to remove the two lead cars from the scene.

Washington's Mayor Adrian Fenty was the dominant public spokesperson in the hours and days following the event. There was some confusion—and resentment—when the mayor, at a press conference, reported the death toll as seven, though investigators and responders had established the count to be nine.[2] Still, the emergency response was efficient and its coordination was attributed in part to improved training since 9/11.[3] Daniel Kaniewski, Deputy Director of the George Washington University Homeland Security Policy Institute, wrote a laudatory assessment in the wake of the crash:

> Yesterday's Metro accident demonstrated that the post-9/11 increased focus on local and regional preparedness has not only better prepared the Washington area for acts of terrorism, but also for the full range of incidents the area faces.... As I monitored the radio traffic of the local agencies involved, I expected to hear chaos; but instead I heard the calm and ordered dispatch of emergency units and informative reports from arriving personnel.... When the DC resources became stretched, pre-identified units from surrounding jurisdictions were alerted and communicated on the same channel as DC units. There were no apparent coordination or communications issues . . . police, fire, emergency medical services, transit, and emergency management officials worked together in a unified manner.[4]

OVERVIEW

Several other high-profile events have also triggered changes in the roles and responsibilities of EMS providers, including the 1993 World Trade Center bombing, the 1995 bombing in Oklahoma City, and then the attacks on September 11, 2001. First responders must now anticipate confronting large-scale incidents with complications not previously planned for. Every EMS provider now needs to anticipate facing an incident that will overwhelm immediately available resources. While not necessarily related to terrorism, or as complex as the events listed above, mass casualty incidents (MCIs) may present common challenges to the responder. This is true whether they involve a 10-patient bus accident or a 500-victim bomb explosion.

Aspects of EMS may vary from one location to another. For example, there are volunteers and career (paid) professionals; municipal, fire department-based, hospital-run, and commercial organizations; and various prehospital provider certifications for first responders including for emergency medical technician (EMT) and Paramedic and Mobile Intensive Care Nurse (MICN). These assorted pieces of the EMS puzzle must assemble, integrate, and collaborate toward achieving the fundamental goal of saving lives. While existing response plans may work within a given service area, when an event threatens to overwhelm local resources, effective state or regional assistance depends on common understandings, interoperability, and coordination. These features must be part of a planning and training program that incorporates experiences and viewpoints from all levels of EMS professionals.

Large-scale Incident

Large-scale incidents may vary in form. Some produce numerous casualties while others, even though requiring a complex response, might result in few or no patients. An act of terrorism would almost certainly qualify as a large-scale incident, either because of mass casualties or because of the greater need for coordination among numerous agencies, or both. An MCI is any event that produces enough patients to potentially overwhelm local resources.[5]

Large-scale incidents and MCIs are generated by various sources. Natural disasters are created by nonhuman forces such as floods, earthquakes, or snowstorms. They may force changes in routine EMS operations by disrupting roads or otherwise hampering communications. Human-caused disasters arise from either conventional or nonconventional methods. Conventional methods, which are most familiar to EMS providers, encompass the "traditional" means of creating numerous patients: motor vehicle accidents, fire, hazardous material release.

Unconventional methods usually refer to the deliberate use of biological, chemical, or radiological agents for criminal or terrorist purposes. (These materials are sometimes referred to as weapons of mass destruction [WMD].) Incidents caused by these agents pose a distinct set of problems for all first responders, not only EMS, as their effects can be widespread and are less familiar. Moreover, their relevance to terrorist or homicidal activity requires close coordination between EMS, intelligence, and law enforcement agencies.

Situational Awareness

After arrival at an event scene, whether a routine medical assignment or a large-scale incident, EMS responders must continuously consider next steps. Maintaining situational awareness and avoiding "tunnel vision" are essential. The training of EMTs and paramedics properly focuses on patient assessment, but it often understates the need for overall awareness of safety risks and other possible impediments to performance. Continual awareness and vigilance after the initial scene size-up is necessary. Potential dangers that EMS providers will encounter can be as simple as slip-and-fall hazards, or as severe as secondary explosive devices (aimed at injuring or killing first responders) or exposure to a WMD. Situational awareness will also help to prevent an easily managed event from progressing out-of-control.

Preparedness

Heightened awareness of the possibility of large-scale incidents has resulted in planning, training, and exercising for EMS providers in conjunction with law enforcement, fire/hazmat, public health departments, and local hospitals.

EMS agencies should undertake strategic planning—developing detailed plans, procedures, and policies to guide responses during routine situations and complex incidents. Planning includes a process of determining the level of threat and hazard, consequences, and vulnerabilities, and then developing strategies to manage risk. Planning documents guide prevention and mitigation efforts and enable effective recovery. A major benefit to the process of developing plans is relationship building. Knowing key individuals as well as the capabilities and limitations of other agencies in advance of an incident will help make an effective response.

The standard EMT and paramedic training curriculum offers minimal information about a provider's additional responsibilities during a large-scale incident.[6] But more information is available from courses on emergency management, and from training specific to the Incident Command System (ICS) and the National Incident Management System

(NIMS). Many states now require ICS and NIMS training for first responders and federal regulations mandate training for the use of incident command in response to hazardous materials events.[7]

EMS responders must complete the training cycle by participating in exercises and drills. Whether in the form of tabletop, functional, or full-scale, exercises enable the testing of plans, protocols, and procedures. They enable first responders to practice, collaborate, and communicate with each other during a simulated event. To fully close the loop, an after-action review—critique of the exercise—is essential to determining what corrective actions might be necessary.

Incident Evaluation

The first EMS unit to arrive at the scene of a large-scale incident has key responsibilities that can determine the course of the event. EMS professionals must immediately evaluate their own safety and the safety of the patients (survivors) and relay their evaluation to a communication center. Beyond the number and severity of casualties, the evaluation should include information about accessibility to the site, safety conditions, and the status of incident management.

Responders must guard against actions that could make conditions worse such as the inclination immediately to care for the patients while not considering the cause of their illness or injury. A memorable example of this was the initial reaction by responders to the 1995 sarin attack in the Tokyo subway. Several had rushed to help victims in contaminated zones but became ill themselves because of failure to first don protective gear.[8,9]

In addition to hazards directly associated with an incident, secondary dangers must also be identified: gas leaks, incoming severe weather, downed power lines, hazardous chemicals, or threats directed at responders such as from improvised explosive devices. Once the scene is deemed safe for rescue operations, EMS should determine/confirm the nature of the incident. The initial assessment provides basic information that guides decisions about further management including what additional resources will be required. Incidents can evolve/dissolve gradually or expand/contract rapidly.

Continuous reevaluation of the scene and of resource allocation is vital to a successful response. EMS workers should always consider the potential for escalating problems. Initially, maintaining a distance from danger zones and having good situational awareness minimize the chances that responders will become patients. Basic safety precaution includes the wearing of personal protective equipment (PPE), establishment of alternative egress routes, and readiness to adapt if an escalation takes place.

Thus, the first unit to arrive at a scene has the responsibility of establishing command, coordination, and communications; initiating triage; and allocating of available resources. Additional triage, treatment, and transport areas can be implemented as follow-on units arrive.

Evidence Preservation

While EMS providers have an overriding objective to provide patient care, when faced with a human-caused incident evidence should be preserved, if possible, for law enforcement authorities. Moreover, the patients/victims themselves potentially can be important witnesses both for medical and legal purposes.

Those who are examined first may be able to provide insight about what can be expected from subsequent patients, possibly influencing operational decisions. In the legal sphere, some victims may have observed the perpetrator and the manner of attack, which could be essential to the criminal investigation.

It is also important to document the positions in which the patients and bodies were found. Their positioning can help determine the size and type of an explosion or identify potential down-wind hazards from toxic materials released at the site. If patients must be moved, or the scene must be disturbed to provide care, a record should be kept of what was moved, when it was moved, where it originally had been, and where it was placed.

TRIAGE

At an MCI, it is important that the first arriving EMS unit implement incident command and begin the triage process. The goal of triage is to do the most good for the most patients and to maximize the number who will survive. Triage in general seeks to ensure that each patient has the opportunity to be assessed, treated, and transported.

The concept of triage seems at odds with a traditional precept taught to EMS workers: provide care for the patient in front of you, especially if someone is critically ill or injured. Triage, on the other hand, avoids prolonged care for any one individual in favor of services for the many, and in the shortest possible time. Even a seasoned EMS professional might be tempted to focus excessively on the most severely injured individual. But concentrating attention and resources on one person deprives others of care that could risk their recovery. In most situations time is of the essence; effective triage can take less than 1 minute to perform on even the most severely injured.[10]

Many patients likely will survive an event with minimal medical care, and others will succumb despite optimal efforts to save them. Some may initially seem minimally hurt but then rapidly deteriorate. Rescuers must try to differentiate between victims who need immediate care and those who can wait. Triaging patients maximizes the use of valuable medical resources and avoids applying "heroic" efforts. By performing triage, rescuers are able to provide a general accounting of all the patients, assess the severity of their injuries, and offer initial care. Using information obtained from the triage process, the appropriate type and quantity of follow-on resources can be requested and tasked effectively.

Simple Triage and Rapid Treatment—START

The Simple Triage and Rapid Treatment (START) system, also referenced in Chapter 4, was developed jointly by Hoag Memorial Hospital and the Fire Department of Newport Beach, California. The system was designed to provide an uncomplicated way to triage patients based on objective physical findings. By establishing standards for triage, START has helped maintain uniformity among different EMS groups. At an MCI, with multiple jurisdictions involved, everyone could then be working from the same standard. Although there are other methods of triage, START has become one of the country's most common approaches.[11]

START utilizes a standard algorithm to categorize or sort patients based on the severity of their injuries, prioritize their need for treatment and transportation, treat the most critical patients first, establish the type and quantity of medical resources needed, and account for the number of patients involved. The system is based on observable patient signs including ability to ambulate, respiratory rate, perfusion, and mental status. Furthermore, it requires minimal medical skills, allows for quick correction of life-threatening conditions while remaining highly efficient, and helps accomplish quick assessment of numerous patients.

The START system places patients into one of four general priority categories:

- Priority 1. Red: Patients who need immediate treatment and/or immediate transport in order to survive
- Priority 2. Yellow: Patients who require medical attention, but will not die or experience extreme adverse consequences if care/transportation is delayed
- Priority 3. Green: Patients who are not critically injured and/or can ambulate and care for themselves; on-scene care and transportation to an acute care facility can be delayed
- Priority 0. Deceased/Black: Patients who show no respiration after one attempt at airway repositioning[12]

Triage is ongoing and does not stop until the last patient has left the transport area. After the first triage sweep, some personnel assigned to this function can be re-tasked. Although reassignment may reduce the number of triage personnel, the reevaluation process must continue. Patients' conditions may change before they are moved to a treatment area, while they are in the treatment area, or while in transit to their final destination. A more thorough assessment in a treatment area may reveal a more serious condition than previously realized. This could require retagging and movement to a higher priority treatment area. Patients may also be moved to a lower priority treatment area if subsequent triage indicates their condition is of lesser concern than initially thought.

JumpSTART is a pediatric-based START system that was created in 1995 to assist the emergency responder in triaging children involved in an incident. It was begun by Dr. Lou Ellen Romig, a Miami pediatric emergency medicine physician. The program utilizes decision points appropriate to the wide variations of normal physiology within the pediatric age group.[13]

Triage Tags

Identifying and tracking patients at a large-scale incident helps providers maintain control of the incident and account for all patients. The triage tag is an important feature toward achieving these goals. A main function of the triage tag is to provide brief information about a patient, including his triage category, specific injuries, and treatment provided at the scene. New Jersey developed its own disaster triage tag based on commercially available models, such as the Medical Emergency Triage Tag (METTAG). The modifications and add-ons designed to enhance the triage process include peel-off stickers; a method to assist in the tracking of a patient's belongings; an overview of the START algorithm; a section for assessing patients who have been exposed to nerve agents; the addition of Chemical-Biological-Radiological-Nuclear-Explosives (CBRNE) contamination symbols.

Triage tags should be stored and packaged in predetermined quantities to promote easy patient accountability by counting used and unused tags when done with the process. As the patient moves through the triage, treatment, and transport system, the triage tag is modified and used to assess and reassess the patient for changes in status. The transporting EMS personnel should record the triage tag number of their patients on each patient's prehospital care report (EMS chart); MCIs do not excuse medical personnel from the duty to complete proper medical documentation on each patient treated and transported.

After arriving at the hospital, the transporting crew should release the patient to the appropriate hospital staff and document the time of the transfer. The crew should be prepared to explain the triage tag to the hospital personnel and advise that the tag become part of the hospital record/patient care report.

Besides the peel-off sticker, another innovation in the New Jersey triage tag structure is inclusion of the "white" category. This category applies to uninjured victims who do not wish to enter into the EMS system. Use of this category allows for an accurate count of individuals involved in the incident while also relieving the EMS system of unnecessary record entries.[14]

Some minimally injured patients might not want to accept on-scene treatment or be taken to the hospital. Even in an MCI, refusal of medical attention (RMA) must be acknowledged with proper documentation provided by medical personnel. Such patients may be asked to remain on scene until all critical patients have been attended to. In some instances, law enforcement or other investigative personnel may wish to speak to patients before they leave the scene or the hospital, so these officials should be informed if a patient desires to leave prematurely.

In any case, reluctant patients should be provided information that accords with local policy so their decision to refuse care will be fully informed. They then should sign the appropriate sections on a patient care report. Upon departure from the scene or hospital, the triage tag should be left behind and kept with the RMA documentation. If a patient were to keep the tag, when patient accountability is later reviewed, the Incident Commander, EMS Branch Director, or other appropriate person in the command structure would not know the disposition of that patient.

COMMUNICATIONS

As demonstrated in many chapters in this book, a common weakness at large-scale incidents is communications failure. Without an adequate communication system, on-scene personnel cannot call for assistance, organize resources, or notify hospitals of casualty numbers and conditions. In addition, without effective communication, an incident cannot be properly managed, which could put responders and members of the public at risk. EMS should work with all potential responding agencies to establish common frequencies or to provide alternative means of information exchange during a response operation. Backup arrangements should be made in any case as a precaution against failure of the primary system.

Communication technology offers a vast array of choices including land-based and mobile telephones, walkie-talkies, computers, tablets, and other digital devices. Connections can be made via voice, e-mail, blogs, or social networks such as Facebook. Each method has advantages and disadvantages. Mobile phones, for example, are portable and can contact distant locations, but are subject to dropped calls. Land-based phones are more reliable, but lack portability.

As part of the planning process, EMS officials must decide which apparatus best fulfills their requirements. Regardless of the type selected, the use of "runners" to hand-carry information from one person to another is a time-tested option, though dependent on available manpower. The type of technology employed at a large-scale incident should be compatible with whatever technology is in everyday use, ensuring that responders are familiar with its operation. When choosing among the various options, other features also need to be considered including flexibility, interoperability, adaptability, and reliability. Additional functional capabilities to be weighed include the length of time before recharging or other service is needed, and what alternate backup system would be most suitable. Failure of entire communication systems—cellular, radio, satellite, Internet—must be considered and planned for.

The use of common terminology is essential in order to avoid confusion among the various personnel at the scene. For example, in various circumstances ambulances have been called buses, trucks, rigs, boxes. If a request for a bus is made, will the response be a transit bus, a school bus, or an ambulance?[15] Using any designation other than the commonly understood one, "ambulance," could cause confusion and error. The same is true for radio call signs such as "EMS Branch Director" rather than "Unit 405." Similarly so for "10-codes," that may be familiar to certain police and fire departments but not to other responders. (New York City Police Department Codes include 10-1 = "call your command," 10-2 = "return to your command," 10-3 = "call dispatcher," etc.[16]) It is better not to use these commands and other terms that might not be familiar outside of a specific jurisdiction, discipline, or system.

ROLES AND RESPONSIBILITIES OF EMS WITHIN AN ICS STRUCTURE

EMS generally falls under the supervision of the Operations Section Chief of the Incident Command System (ICS) and commonly is placed at the branch level (an intermediate organizational step within the section). The EMS Branch Director is the person in charge of all EMS operations at an incident and will designate group supervisors and other assignments based on the needs, size, and scope of the event. Within the EMS Branch, additional organizational levels are normally established for the functions of triage, treatment, transport, and a possible separate EMS Staging Area.

The Triage Group is responsible for performing all primary and secondary triage of patients. The Treatment Division handles all on-scene treatment of patients until transport is initiated. Depending on the number of patients (patient load) and the available transport resources, the Treatment Division may be divided into three priority areas designated by the color coding (red, yellow, green) as described earlier in the section on START.[17]

The fourth priority area, black, generally is not established by EMS; the medical examiner or other law enforcement personnel is responsible for this task that includes the establishment of morgue facilities. The deceased individuals should not be moved until the viable patients have been triaged and treatments initiated.

The Transport Group/Division is responsible for coordinating all patients' transportation from the scene to medical facilities via ground ambulances, air-medical transport resources, and other vehicles such as transit or school buses. (Ambulances are variously equipped with Advanced Life Support or Basic Life Support capabilities, designated as ALS or BLS.) The Transport Group/Division Supervisor has the responsibility for distribution of the patient load to local hospitals and specialty facilities, such as trauma and burn centers.

Depending on the scale of the incident and the resources needed, the Operations Section Chief may designate a separate staging area for EMS vehicles. Since EMS units will likely be entering and departing the scene frequently, it may be practical to have a staging area explicitly for EMS vehicles. The EMS vehicles should be stationed distinctively in ALS and BLS sections.

In the EMS Staging Area, units may unload extra supplies needed for the treatment areas, as directed by the EMS Branch Director or the Staging Area Manager. Some systems suggest that stretchers be removed from the individual units while in staging so that a patient can be loaded on the stretcher while still in the treatment area. One pitfall of this practice is that not all ambulance stretcher mounts are compatible with every type of stretcher, and compatibility problems will slow down transport activities. The driver of each vehicle should remain with the vehicle so it can be moved immediately.

Some systems will have EMT personnel exit the ambulance and assist with scene operations. Then, as transports are initiated, the first available EMT or paramedic will transport that patient in the next available ambulance from the staging area. But switching personnel, like switching stretchers, risks confusion and reduction in efficiency. Some of the first arriving EMS units may not be available for transport because the personnel and equipment from those units have been deployed to the treatment, transport, or triage areas. These units should be parked out of the way beyond the service area to avoid interference with in-service units.

Although most responsibilities of EMS within the ICS structure will be in the Operations Section, at some incidents a Medical Unit within the Logistics Section will also be needed. While the Operations Section provides medical treatment to victims of the incident, the purpose of the Medical Unit is to develop a plan and provide care and support to incident personnel. Depending on the scope and nature of the incident there may also be other EMS-specific roles within the ICS structure, such as Supply Unit or Safety Officer.

RECOVERY FROM THE INCIDENT

There are several issues to address as an incident draws to a close. Of primary concern is the health of personnel and the loss, damage, or contamination of equipment. Responders involved in the event may have developed physical or mental health issues as a result of their participation and exposure. Equipment must be evaluated to ensure it is still operational. Questions such as "how hard was it used?" and "how much longer will it last?" should be considered.

Following the incident and debriefing of personnel, especially those with major response roles, an after-action report should be produced quickly. Protocols may then need adjustment in accordance with lessons gleaned from the incident.

The mental health of responders is a key issue that needs to be addressed.[18] The use of Critical Incident Stress Management (CISM) programs can assist EMS and other responder personnel to cope with their current work, the next large-scale incident, and family and personal responsibilities. Much like the definition of an MCI, a critical incident is any event that has the potential easily to overcome a person's normal ability to cope with stress. Techniques and programs include defusing (speaking with a trained professional on site, or as soon as possible after the incident); debriefing (similar to defusing; normally in a group setting); and other individual and group therapy and mental health treatment.

These are designed to help the responders deal with physical or psychological symptoms associated with exposure to an event; they enable ventilation of emotions, offer reassurance and support, decrease the potential for the development of psychological problems, and allow for identification of those individuals who may require additional help.

CONCLUDING OBSERVATIONS

In the late 1990s and early 2000s, a gap in training of EMS professionals was identified in New Jersey by the Center for BioDefense at UMDNJ-New Jersey Medical School. While there was ample training on specific topics such as ICS, medical treatment, and hazardous materials, no course for EMS workers had "put it all together." As a result, novel training programs grew out of efforts at the Center, along with support from the state Department of Health and other EMS and health officials. Two courses were established on "EMS Response to the Large Scale Incident." One focused on awareness and the other on operations, though both tied together key concepts. Many areas discussed in this chapter are covered in these two courses, at a basic level (awareness), and more in depth (operations).

Another outgrowth of these courses, the Center's programs, and New Jersey's EMS community, is the New Jersey EMS Task Force (NJEMSTF). The NJEMSTF was established in 2003 to help support large-scale, complex, and other special situations that require personnel, equipment, and materials not traditionally available to EMS agencies. The NJEMSTF has trained members to staff staging areas, helicopter bases, medical units, EMS branches, and regionally based vehicles to respond to high-impact incidents.[19] Composed of more than 250 volunteer EMTs, paramedics, nurses, physicians, communicators, and other support staff, the NJEMSTF may be seen as a model organization for in-state and out-of-state emergency medical missions.

NOTES

1. National Transportation Safety Board. Collision of Two Washington Metropolitan Area Transit Authority Metrorail Trains Near Fort Totten Station, June 22, 2009. Railroad Accident Report. Washington, DC. Adopted July 27, 2010. Available at www.ntsb.gov/doclib/reports/2010/RAR1002.pdf

2. Cherkis J. So who screwed up the Metro crash body count? Fenty. Washington City Paper, June 24, 2009. Available at http://www.washingtoncitypaper.com/blogs/citydesk/2009/06/24/so-who-screwed-up-the-metro-crash-body-count-fenty/, accessed June 18, 2012.

3. Cook D. Emergency response to Metrorail crash shows post-9/11 gains. The Christian Science Monitor, June 24, 2009. Available at http://www.csmonitor.com/USA/2009/0624/p02s21-usgn.html, accessed June 18, 2012.

4. Kaniewski DJ. The Metrorail crash: An effective regional response. Homeland Security Policy Institute, Washington, DC, June 23, 2009. Available at www.gwumc.edu/hspi/policy/commentary04_MetroResponse.cfm, accessed June 18, 2012.

5. Mass casualty incident: An overview. Available at http://www.emsconedonline.com/pdfs/EMT-Mass%20Casualty%20Incident-an%20overview-Trauma.pdf, accessed February 20, 2012.

6. US Department of Transportation. EMT-Basic: National Standard Curriculum. Walt A. Stoy, Principal Investigator, University of Pittsburgh, School of Medicine, 2009. Available at www.nhtsa.gov/people/injury/ems/pub/emtbnsc.pdf, accessed June 18, 2012.

7. US Department of Homeland Security, Federal Emergency Management Agency. Incident Command System. Available at http://www.fema.gov/emergency/nims/IncidentCommandSystem.shtm, accessed June 18, 2012.

8. Cole LA. *The Eleventh Plague: The Politics of Biological and Chemical Warfare.* New York: WH Freeman; 1997. pp 151–153.

9. Pangi R. Consequence management in the 1995 sarin attacks on the Japanese subway system. Stud Confl Terrorism 2002;25:432–448.

10. Simple Triage and Rapid Treatment (START). Community Emergency Response Team, Los Angeles. Available at http://www.cert-la.com/triage/start.htm, accessed June 18, 2012.

11. US Department of Health and Human Services, Chemical Hazards Emergency Medical Management. START Adult Triage Algorithm. Available at http://chemm.nlm.nih.gov/startadult.htm, accessed June 18, 2012.

12. Ibid.

13. Romig LE. The JumpSTART Pediatric MCI Triage Tool. Available at http://www.jumpstarttriage.com, accessed June 18, 2012.

14. NJ Department of Health and Senior Services, Office of Emergency Medical Services. NJ Disaster Triag Tag. Available at http://www.nj.gov/health/ems/documents/njdisastertag.pdf, accessed June 18, 2012.

15. EMTLife, Online Forum for EMS-Related Discussion. What do you call an ambulance? Available at http://www.emtlife.com/showthread.php?t=6964, accessed June 18, 2012.

16. New York City Police Department. Radio signal codes. Available at http://www.n2nov.net/nypdcodes.html, accessed June 18, 2012.

17. US Department of Homeland Security, Federal Emergency Management Agency. Incident Command System.

18. Centers for Disease Control and Prevention, Emergency Preparedness and Response. Disaster Mental Health for Responders: Key Principles, Issues and Questions. Available at http://www.bt.cdc.gov/mentalhealth/responders.asp, accessed June 18, 2012.

19. NJ Department of Health and Senior Services, NJ Emergency Medical Services Task Force. Available at http://www.nj.gov/health/ems/njemstf.shtml, accessed June 18, 2012.

8

THE ROLE OF THE MENTAL HEALTH PROFESSIONAL

Ann E. Norwood, Lisa M. Brown, and Gerard A. Jacobs

This chapter discusses the role of mental health professionals in terror and disaster response drawing heavily from the 9/11 World Trade Center (WTC) attacks to illustrate concepts relevant to immediate interventions. After a brief description of the attack, the chapter focuses on elements that have particular salience for behavioral healthcare providers, gives examples of immediate mental health services provided at the time and progress made since then, and ends with a discussion of current and future contributions that behavioral health professionals and researchers are poised to make to disaster care.

THE 9/11 ATTACKS

On September 11, 2001, a sunny, crisp autumn day in Lower Manhattan was shattered at 8:46 A.M. when American Airlines Flight 11 crashed into the North Tower of the WTC. The crash of a second aircraft, United Airlines Flight 175, into the South Tower at 9:03 A.M. removed any uncertainties that the destruction was accidental. During the 102 minutes between the first aircraft strike and the collapse of the North Tower at 10:28 A.M., millions around the world followed the unfolding tragedy on TV, the Internet, and radio. Broadcast coverage included images of people leaping to their deaths, terrified survivors caked in dust and debris running for their lives, and endless replays of the planes striking the towers. Destruction of the WTC led to power outages and the loss of cell towers and phone lines, combined with a surge in demand, severely disrupted telephone communications. Over hours and days, family and friends of the missing posted photographs of their loved ones in hopes of learning their fate.

The event remains sharply etched in the memory of Americans and, for many, returned anew with the remembrance ceremonies on its 10th anniversary. In the wake of 9/11, much additional knowledge has been learned about individual and community psychological and

Local Planning for Terror and Disaster: From Bioterrorism to Earthquakes, First Edition.
Edited by Leonard A. Cole and Nancy D. Connell.
© 2012 John Wiley & Sons, Inc. Published 2012 by John Wiley & Sons, Inc.

social responses to traumatic events. However, while the use of evidence-based practices to treat psychological symptoms shows great promise in improving survivors' mental health outcomes,[1,2] most people who would benefit from intervention do not receive services.[3–5] Evaluations of disaster mental health programs reveal that only a small proportion of people who reported significant symptoms received care during the recovery phase.[6–8]

In marked contrast to the wealth of studies examining the mental health consequences of disasters,[9,10] comparatively little attention has been focused on identifying methods to enhance acceptance of disaster mental health services other than to describe the strengths and weaknesses in the current delivery systems. As a result of these challenges, researchers, program evaluators, and policy makers are interested in studies examining the effectiveness of psychological first aid (PFA) and other interventions intended for disaster-exposed populations, exploring why people in need of intervention do not seek or accept available treatment, and developing new approaches to outreach and delivery of disaster behavioral health services. The need for efficacious interventions that are readily and easily implementable is great and growing.

EARLY MENTAL HEALTH RESPONSES

As it is for other healthcare providers, lifesaving is a top priority for mental health professionals following a disaster. Hospital-based mental health professionals are often among the first to be involved in response-and-support lifesaving activities. Following a disaster, hospitals are frequently besieged by non-critically injured people—including members of the media and those looking for loved ones or shelter. Mental health providers can maximize lifesaving by helping divert non-emergent persons to preserve critical access for those in urgent need of care. The following descriptions of hospital-based 9/11 activities illustrate some of the roles that mental health professionals performed.

Hospital-Based Mental Health Responses

Within minutes of the first plane striking the North Tower, New York City (NYC) hospitals were placed on alert. Across the city, psychiatry departments joined others to prepare to care for casualties. As is common following disasters, many people converged on hospitals for reasons other than injuries. They went to donate blood or volunteer in some other capacity, look for loved ones who might have been injured, or seek a safe refuge that had water, electricity, and food.

New York University (NYU) Downtown Hospital, located four blocks from the WTC, began receiving hundreds of patients shortly after the collapse of the South Tower.[11] By mid-afternoon, survivors stopped arriving at the hospital. As the anticipated second wave of trauma patients failed to materialize, the painful realization that there were few survivors became evident at hospitals across the city:

> We could not believe it. . . . That it could be damaged I could accept, but when I learned that the Towers had collapsed, I was just speechless. I could not believe it. . . . You could not even begin to think about the human toll at first, inasmuch as you were trying to respond to the situation itself, which was so shocking. . . . It turned out that there was no need because there were no survivors of the magnitude we anticipated. That was both surprising and horrifying as we began to understand why.[12]

In the days that followed, NYU Downtown Hospital was one of few public institutions operating in Lower Manhattan. The hospital undertook a number of initiatives to take care of the basic needs of community residents including outreach to residential facilities to ensure the availability of medications and provision of meals from the hospital cafeteria.

At St. Vincent's Hospital, the level I trauma facility closest to the WTC, the Emergency Department received 400 patients in the first 2 hours, most with minor injuries.[13] Despite the paucity of patients, however, the hospital was soon inundated with individuals volunteering to help and with people searching for missing loved ones. The hospital established a family center but was quickly overwhelmed with anxious and grieving people. Located two blocks away, the New School University offered space for the hospital to open a second family center. By evening, both family centers had hundreds of people seeking information or counseling waiting to enter.[14] Over the course of that week, staff at the family centers provided direct services to more than 7000 people in person and more than 10,000 people by phone.[15]

At Columbia Presbyterian Medical Center on the Upper West Side, mental health professionals prepared information for distribution throughout the hospital about common emotional responses to a disaster. They also created and staffed four information centers.[16] In addition they established a 24-hour hotline for disaster-related calls and staffed a station near the emergency department to render assistance. On September 13, a town hall for all mental health employees included education on disaster responses by experts in the field, discussions of how to support one another, and how to identify and refer those who might need professional assistance.[17]

Public Mental Health Response and Consultation

At the time of the attack, Dr. Neal Cohen was commissioner of both the NYC Department of Health and Department of Mental Health. As a psychiatrist, he was uniquely positioned to advise Mayor Rudolph Giuliani, shape the public mental health response, and ensure its coordination and integration with physical health. Cohen raced to the scene to join the mayor and was at his side throughout the day to provide consultation. At the mayor's request, Cohen brought in experts in bereavement and traumatic stress to help inform the mayor about respecting the grief of those whose loved ones died and promoting community resilience in his communications with the public.

The commissioner described three main goals of the public mental health response[18]:

1. Prioritize populations into higher- and lower-risk categories based on the directness of the WTC impact upon them (e.g., survivors, victims' family members, eyewitnesses, residents of Lower Manhattan).
2. Mobilize the social support networks of surviving coworkers and victims' families.
3. Use healing and help-promoting messages as part of a public education and community outreach since all New Yorkers were affected to some degree.

Community-Based Organizational Responses in the Immediate Aftermath

Initial community mental health interventions were provided by existing organizations with a focus on mental health and survivor assistance. Many of these organizations were directly affected by the WTC attack. Some had office space in Lower Manhattan or adjacent neighborhoods; staff members had loved ones who worked in areas in harm's way; the closure of bridges and tunnels precluded some from getting to work; and the disruption of communications interfered with established hotlines. Examples of the experience of two major organizations that provide mental health services are described in the following sections.

LifeNet   LifeNet, a 24/7-multilingual NYC hotline staffed by mental health professionals, provides a crisis information and referral system for persons requesting help for psychological or substance abuse. It was established in 1996 by the city's Department of Mental Health (DMH) under a contract with the Mental Health Association of NYC. On 9/11, it had a staff of approximately 50 seasoned professionals. Their toll-free phone line was dead after the towers collapsed. They quickly contacted the media to publicize an alternative phone number. Their first 9/11-related distress call came that night from a man who wanted information about where he could find a lost loved one. Otherwise, the phones were "chillingly quiet."[19] Initial calls consisted primarily of people offering their assistance at the City's Family Assistance Center and, in a first for LifeNet, requests for counseling services by schools, workplaces, and community centers.

LifeNet received its first 9/11 mental health call on September 12. A delivery truck driver sought a counselor to help him with the emotional impact of what he experienced the day before. On that morning he had just completed a delivery to the North Tower's 110th floor when he heard the first plane hit. It took him 20 minutes to make his way down the stairs to his truck and by then the second plane hit. He had to drive his truck over rubble and human remains in order to leave the area.

Safe Horizon   Safe Horizon is the largest victim assistance and advocacy organization in the NYC area. Founded in 1978, it serves victims of crime and abuse, offering a number of services including crisis intervention, counseling, practical support, information, education and referrals, and hotlines. Their administrative headquarter is located six blocks from the WTC complex (and some staff were eyewitnesses to the attack); their other 80 offices are scattered throughout the city. On 9/11, they had a staff of more than 900 employees. Like LifeNet, Safe Horizon's hotline was used extensively. They began receiving hundreds of calls a day from people seeking information on a broad range of topics including how to find missing loved ones, where to look for housing following displacement, what to do about displaced jobs, and how to cope with the fear of further terrorist attacks.[20] Their immediate focus was on crisis intervention and services to strengthen people's ability to cope with traumatic stress.

DISASTER MENTAL HEALTH

Psychological Responses to Disaster

Disasters elicit a range of psychological responses that are influenced by intrapersonal factors and interpersonal circumstances, as well as by the nature and characteristics of the event.[21] Decades of research and experience have shown that distress is virtually universal for those affected by a disaster but that for most people, it resolves over time without professional intervention. For some, however, adverse responses and impairment in functioning persist and may even become psychiatric disorders. The most common negative psychological outcomes from disaster exposure are anxiety, depression, post-traumatic stress disorder (PTSD), nonspecific psychological distress, and health concerns.[22] Although a relatively small number of people may experience enduring symptoms, numerous studies have documented the adverse consequences of untreated depression, anxiety, and PTSD.[23-25] The 9/11 attacks and other disasters in the last decade have highlighted the importance of psychological resilience and time-phased interventions to promote it.

How Disaster Behavioral Healthcare Differs from Traditional Behavioral Health Intervention

Professionals from many disciplines are involved in behavioral healthcare following disasters, including psychiatrists, psychologists, counselors, social workers, marriage and family therapists, and psychiatric nurses. Disaster mental healthcare differs in many ways from traditional mental healthcare. It emphasizes people experiencing usual responses to extraordinary events rather than patients experiencing pathology. Traditional mental health treatment differs from disaster mental health in that the former "implies the provision of assistance to individuals for an existing pathological condition or disorder," whereas the latter assumes "that the individual is capable of resuming a productive and fulfilling life following the disaster experience if given support, assistance, and information at a time and in a manner appropriate to his or her experience, education, developmental stage and culture."[26]

Disaster survivors who use disaster mental health services have been called "accidental clients," because an external event as opposed to an intra- or interpersonal issue resulted in their use of services.[27] Accordingly, people who would benefit from disaster mental health services may not self-identify as being mentally ill or having a mental health problem and thus refrain from accepting treatment, especially if offered through a mental health clinic. The initial focus of disaster mental healthcare is on groups of people (populations) rather than individuals. Outreach is used extensively rather than waiting for patients/clients to come to offices. Disaster mental health professionals must also learn to work closely with disciplines with which they usually have little contact such as public health, public safety, and emergency management.

Moreover, most people with serious psychiatric illness prior to a disaster share few characteristics with those without a history of mental illness who are currently experiencing disaster-related distress.[28] Because the goal of disaster mental health intervention is to help survivors return to their pre-disaster level of functioning, the level of treatment offered by crisis counseling may not be optimally therapeutic for people with serious psychiatric illness. The existing literature indicates that most people with well-controlled mental illness prior to a disaster function fairly well during the recovery phase, whereas those with mental illness that was not adequately managed are at risk for experiencing an exacerbation of psychiatric symptoms, especially if essential services were interrupted.[29] For these latter individuals, the goal generally is to restore essential services and to refer them to local mental health providers who can offer ongoing psychiatric care.

Terrorism differs from natural disasters and industrial accidents by virtue of the fact that it is intentional. A sense of vulnerability may be generalized from fear that further attacks could happen anywhere. Confidence in public institutions drops because of their failure to protect citizens. Moreover, fear and anger can lead to scapegoating of innocents with ethnic or other characteristics similar to those of perpetrators.

Education and Training

To be effective care providers following a disaster, focused education and training are important for mental health professionals. As noted earlier, disaster work requires reorientation from usual practice for most mental health professionals. Fortunately, most professional organizations are incorporating opportunities to learn more about disaster behavioral health into their meetings and continuing professional education.

In the immediate aftermath of the attack on the WTC, many mental and behavioral health professionals (MBHPs) stepped forward and offered their services in the disaster relief operation. Many of them were insistent that they *had* to be a part of the response. The great majority of those professionals did not have training in disaster mental and behavioral health, or even traumatic stress. Similar cadres of event-specific volunteer MBHPs appear at every high-profile disaster. There is broad appeal to being part of the solution, but doing the preparation necessary to be effective in such situations does not seem to be as alluring. Such preparation is, however, essential and accessible. MBHPs need to reorient as described earlier and need to acquire topical knowledge about disaster and traumatic stress.

The National Biodefense Science Board (NBSB) in its 2008 report on disaster mental health preparedness recommended that training in disaster mental health become a standard requirement for graduate training in the mental health professions.[30] The American Red Cross has offered free courses for MBHPs since 1992. The training period has become progressively shorter, beginning as a 2-day training, and devolving to its current 3-hour incarnation. But the Red Cross training focuses on how to use one's professional skills while working under the auspices of the Red Cross, rather than actually providing training in disaster mental health. Participants have always been encouraged to obtain training in the field outside the Red Cross course. Various MBHP guilds have offered training in disaster mental and behavioral health at times, but none have a sustained program of training such as that recommended by the NBSB.

The Disaster Mental Health Institute (DMHI) at the University of South Dakota has offered a Graduate Certificate in Disaster Mental Health since 1997. Beginning in 2009 the courses are all offered online, and the eligibility was expanded to include the groups encouraged by the NBSB recommendations to obtain such training. Required courses include Disaster Mental Health, Crisis Intervention, and Serving the Diverse Community in Disaster. A fourth course is chosen from electives including Traumatic Stress and Management in Disaster Mental Health. The DMHI also offers a Doctoral Specialty Track in Clinical/Disaster Psychology. A variety of other universities offer occasional courses on these topics.

DISASTER BEHAVIORAL HEALTH INTERVENTIONS

Identifying Groups at High Risk for Negative Psychological Outcomes

One of the early tasks of behavioral healthcare provision is delineating the traumatic stressors associated with the disaster. While each disaster has unique elements, many stressors are common across disasters. Using this knowledge, planners and practitioners then try to identify groups that have been exposed to them in order to prioritize and tailor service delivery and outreach efforts. In assessing needs and planning interventions, disaster behavioral health experts first attempt to characterize the risks posed by the traumatic stressors and then think through the populations that were likely exposed to them.

The level of exposure to traumatic stressors is the strongest predictor of future psychological and social impairment.[31] However, 9/11 experiences suggest that direct exposure alone does not determine the degree to which people have strong psychological responses.

TABLE 8.1   Traumatic Elements Associated with 9/11

Traumatic Elements of Attack on World Trade Center
Large loss of life
Loss of work groups
Traumatic images (e.g., people jumping from Towers)
Limited communications
Collapse of structures/dust clouds
Novel use of aircraft as explosive devices

Post-attack Traumatic Elements
Recovery of human remains
Lack of remains for many people
Loss of loved ones
Injuries
Dislocation
Job loss
Fear of additional attacks
Potential exposure to hazards of dust and debris throughout Lower Manhattan

The infliction of intentional harm has a more harmful psychic valence than accidents or natural disasters. Thus, the fact that terrorists deliberately flew the planes into the WTC Towers increased the "dose" of traumatic stress. Additional factors that amplified the psychological effects included the novel use of aircraft as explosive devices, the destruction of iconic buildings, the coordinated nature of the attacks, and their duration. In contrast to the 1995 bombing of the Alfred P. Murrah Federal Building in Oklahoma City, which immediately brought down most of the building, the WTC attacks were protracted. Over the span of 102 minutes, the first plane struck the North Tower, the second plane hit the South Tower, the South Tower collapsed, and finally the North Tower collapsed.[32] The drawn-out tragedy was broadcast live and replayed endlessly on television and other media. Additional intra- and post-event factors were traumatizing and would be noted by disaster mental health providers. These are summarized in Table 8.1.

Family, Friends, and Coworkers of the Dead

The magnitude of a disaster is often characterized by the number of people killed and injured. These same figures help predict the scale of associated psychiatric morbidity. Of the estimated 17,400 occupants inside the towers at the moment the first plane hit, 2163–2180 perished.[33] A total of 2752 people (other than terrorists) including emergency responders, bystanders, and airline passengers and crew were killed or reported missing and believed to be dead.[34]

In planning for psychological interventions and prioritizing high-risk groups, it is also important to note that many work groups sustained large losses of life and may therefore be at greater risk. For example, those who died included 343 firefighters and paramedics; 60 NYPD officers and Port Authority police officers; 658 employees of the financial services firm Cantor Fitzgerald; and people from 60 companies.[35] Other high-risk groups include those whose spouses or partners died (1609) and children who had a parent die (3051).[36] In NYC, 11–14% of adults reported losing a friend or relative, and the national level studies reported that 4–11% of the US adult population knew someone who was killed.[37–39] People around the world were also touched by the loss. Citizens from 115 nations died.[40] Loved ones of those who die suddenly and violently in traumatic events will experience grief and, for some, the development of "complicated" or "traumatic" grief characterized by chronic yearning, mourning and loss-related withdrawal and anguish.[41] Furthermore, field experience strongly suggests that the death of a loved one in a situation where many others die is likely to exacerbate the grief.

The Injured

Another group at higher risk for psychiatric sequelae includes those injured during a disaster. One indicator of the number of injured includes those presenting to emergency departments for evaluation and treatment. An early study of the four hospitals closest to the WTC complex and a fifth hospital that served as a referral center for burned patients documented a total of 790 survivors with injuries within the first 48 hours. Among these survivors, 386 (49%) had inhalation injuries from smoke, debris, dust, or fumes, and 204 (26%) had eye injuries.[42] One hundred thirty-nine patients were hospitalized.[43] However, this proved to be the proverbial tip of the iceberg. In the months and years that followed, more and more people sought attention for medical conditions potentially related to health effects of exposure to toxic substances in the dust and debris.[44]

Those Who Experienced Threat to Life

WTC survivors of the attack experienced many life-threatening elements. These included the physical impact of the planes striking the towers, fire and smoke, impaired egress, the need to evacuate via stairs (especially for those with disabilities), loss of power, and confusion about what had transpired. As the towers collapsed, the clouds of dust and debris and the attempt to escape them were terrifying, and many reported that they thought they were going to die.

Those Who Witnessed Horrific Events

The juxtaposition of a beautiful autumn day with jetliners crashing into a prominent landmark occupied by thousands of people stunned both those witnessing the events in person and those watching on TV. For many, horrifying images lingered long afterward: the planes striking the towers, people jumping from windows to escape fire, the towers' collapse, faces displaying terror and grief, people trying to outrun the dust cloud. These images added to the challenge of progressing from the initial response efforts toward recovery.

POST-DISASTER STRESSORS AFFECTING PSYCHOSOCIAL OUTCOMES

The development of psychiatric disorders is not determined solely by a disaster experience. Pre-disaster elements such as pre-existing psychiatric illness, poor social support, concurrent stressors, and socioeconomic status can influence who becomes ill. Similarly, the post-event environment plays a role in determining mental health outcomes; ineffective disaster relief efforts, relocation, job loss, and other adversities can contribute to impaired functioning and mental illness.

Moreover, in the case of the WTC, there continued to be a series of traumatic stressors for those involved in recovery efforts. For example, the identification of human remains is psychologically (and often, physically) taxing. Sorting through rubble for missing bodies and body parts, especially with the intense heat of the pile and the clouds of smoke in the initial weeks, going through personal effects, asking families for physical descriptions of lost loved ones, conducting death notifications, and other related activities are emotionally (and in some cases, physically) demanding. Moreover, many of those involved in response activities may later develop medical illnesses related to toxic environmental exposures, adding another set of psychological and physiological demands.

Job Loss

Post-disaster resource losses negatively affect people's ability to recover psychologically. At its peak, job loss in NYC attributed to the attacks was estimated at between 49,000 and 71,000.[45] Effects were most pronounced in the area surrounding the WTC where many stores, offices, and businesses were destroyed or severely damaged. Large employment effects extended throughout Lower Manhattan where access was difficult due to damaged public transportation infrastructure. Drops in tourism and cascading effects from the loss of financial sector jobs contributed to business losses throughout the city.[46] Most of the job loss was centered in the airline, finance, hotel, and restaurant industries.

Ongoing Adversities

Many people who resided in the area surrounding the WTC complex were temporarily or permanently displaced because of building damage or dust abatement in their homes. Both displacement and relocation are associated with increased risk for psychiatric morbidity.

RANGE OF EARLY DISASTER BEHAVIORAL HEALTH INTERVENTIONS

Community-based PFA is an intervention strategy that focuses on teaching basic psychological support to the general public, who can then provide better support to family, friends, and neighbors, as well as handle stress more effectively themselves.[47] The community-based model molds the training to fit the worldview, ethnic and cultural traditions, and other features of the targeted community. The training can also be adapted for specialized populations such as first responders, health professionals, and civic leaders.

Typical modules in the training include learning how to reach out to help people, understanding traumatic stress, active listening, coping and problem solving, grief and bereavement, when to make referrals, self-care, and ethical considerations. Some communities also choose to add a component focusing on the needs of children.

Community-based PFA was chosen in 1995 by the International Federation of Red Cross and Red Crescent Societies for helping developing countries begin psychological support programs. In 2002, the American Red Cross asked their "technical experts" in disaster mental health to examine the responses to the terrorist attacks of September 11 and suggest changes to the extant Disaster Mental Health model. The technical experts unanimously agreed that it would be profitable to add community-based PFA in addition to the Disaster Mental Health model both for Red Cross relief workers and for the public. The Institute of Medicine's (IOM) Committee on Responding to the Psychological Consequences of

Terrorism made part of its first recommendation that education in PFA be developed for the general public.[48] Furthermore, the NBSB recommended to the Department of Health and Human Services that civic leaders, first responders, and the public be trained in PFA.[49] However, it is important to note that both the IOM committee and the NBSB noted that more research was needed on the effectiveness of community-based PFA.

Leaders in the private and public sector may find it helpful to request consultation from mental health experts. Understanding how people respond to a disaster can inform policies and communications that help survivors recover psychologically.

Mental health professionals, who are licensed to prescribe, can help replace lost psychotropic medications. Some MHPs can also assist in diagnosing medical conditions that present with unusual behavior or perceptions.

LESSONS LEARNED: THE NEED FOR EFFECTIVE OUTREACH AND MEDIA CAMPAIGNS

Reviews of mental health crisis counseling programs conducted in the aftermath of 9/11 and other disasters reveal that significant challenges are encountered by agencies responsible for the delivery of crisis counseling and disaster mental health services.[50–52] In general, these programs were not reaching many of the people they were intended to serve.

To encourage use of services, after the events of September 11, 2001, the New York State Office of Mental Health initiated a media campaign to educate the public about available counseling programs. The campaign was launched because it was recognized that mental illnesses are often stigmatized and people do not anticipate needing mental health treatment after a disaster. As a result of the media campaign, counseling program awareness increased from 30% approximately 3 months after the attacks to 50% at the 1-year anniversary. Of the 50%, 30% reported that they considered contacting or had contacted a mental health service provider.[53]

Knowledge of the most effective messages and the best modalities for delivery of messages is vital to creating awareness and educating those who could potentially benefit from services, but who do not possess fundamental knowledge about how to access the system or understand the value of disaster mental health services. In addition, clinicians responsible for providing services may not possess marketing knowledge and may already be stretched thin in meeting the needs of their clients. Historically, funding agencies have not provided sufficient or dedicated monies to hire a person to conduct targeted outreach.

CONCLUSION

Disaster mental health practitioners can make important contributions in a disaster. To do so, however, takes preparation. To be most effective, it is incumbent on the mental health professional to

- receive special training in disaster mental health and in general principles of disaster planning and response
- become affiliated and credentialed with an organization involved in disaster response and/or with an institution, who may wish to consult with you before, during, and/or after a disaster

NOTES

1. Hamblen JL, Gibson LE, Mueser KT, Norris F. Cognitive behavioral therapy for prolonged postdisaster distress. J Clin Psychol 2006;62:1043–1052.

2. Wilson JP, Friedman MJ, Lindy JD, editors. *Treating Psychological Trauma and PTSD*. New York: The Guilford Press; 2001.

3. Wang PS, Gruber MJ, Powers RE, et al. Mental health service use among Hurricane Katrina survivors in the eight months after the disaster. Psychiatr Serv 2007;58(11):1403–1411.

4. Wang PS, Gruber MJ, Powers RE, et al. Disruption of existing mental health treatments and failure to initiate new treatment after Hurricane Katrina. Am J Psychiatry 2008;165(1):34–41.

5. Galea S, Brewin CR, Gruber M, et al. Exposure to hurricane-related stressors and mental illness after Hurricane Katrina. Arch Gen Psychiatry 2007;64(12):1427–1434.

6. Ibid.

7. Wang, 2008.

8. Wang, 2007.

9. Norris FH, Friedman MJ, Watson PJ, et al. 60,000 disaster victims speak: Part I. An empirical review of the empirical literature, 1981–2001. Psychiatry 2002;65(3):207–239.

10. Ibid., Part II. Summary and implications of disaster mental health research. pp 240–260.

11. Wang W. The hospital in its community. In: Danieli Y, Dingman RL, editors. *On the Ground after September 11: Mental Health Responses and Practical Knowledge Gained*. Binghampton, NY: The Hayworth Press; 2005. pp 19–23.

12. Oldham JM. September 11, 2001 and its aftermath, in New York City. In: Ursano RJ, Fullerton CS, Norwood AE, editors. *Terrorism and Disaster: Individual and Community Mental Health Interventions*. Cambridge, UK: Cambridge University Press; 2003. pp 23–30.

13. Eth S, Sabor S. Healing in the aftermath of 9/11: Recovery from suffering and grief for the community and its caregivers. In: Danieli Y, Dingman RL, editors. pp 42–50.

14. Ibid.

15. Ibid.

16. Oldham.

17. Ibid.

18. Cohen NL. Reflections on the public health and mental health response to 9/11. In: Danieli Y, Dingman RL, editors. p 26.

19. Draper J. LifeNet and 9/11: The central role. In: Danieli Y, Dingman RL, editors. pp 63–82.

20. Arnow N. Safe horizon's response to 9/11: Reflections on the past and a renewed focus on the future. In: Danieli Y, Dingman RL, editors. pp 56–71.

21. Norris, Friedman, Watson, et al.

22. Ibid.

23. Amaya-Jackson L, Davidson JR, Hughes DC, et al. Functional impairment and utilization of services associated with posttraumatic stress in the community. J Trauma Stress 1999;12(4):709–724.

24. Kessler RC. Posttraumatic stress disorder: The burden to the individual and to society. J Clin Psychiatry 2000;61 Suppl 5:4–12.

25. Weisler RH, Barbee JG 4th, Townsend MH. Mental health and recovery in the Gulf Coast after Hurricanes Katrina and Rita. JAMA 2006;296(5):585–588.

26. Federal Emergency Management Agency [FEMA]. FEMA Fact Sheet: Crisis Counseling Assistance and Training Program. Available at http://www.fema.gov/pdf/media/factsheets/2009/dad_crisis_counseling.pdf, accessed June 18, 2012.

27. Silk S. Preparing psychologists for the reality of disaster work: Distinctions between traditional psychotherapy and disaster mental health. Paper read at 114th Annual Convention of the American Psychological Association, New Orleans, Louisiana, 2006.

28. Katz CL, Pellegrino L, Pandya A, et al. Research on psychiatric outcomes and interventions subsequent to disasters: A review of the literature. Psychiatry Res 2002; 110: 201–217.

29. Ibid.

30. Disaster Mental Health Recommendations: Report of the Disaster Mental Health Subcommittee of the National Biodefense Science Board. Approved by the NBSB November 18, 2008. Available at http://www.phe.gov/Preparedness/legal/boards/nbsb/Documents/nsbs-dmhreport-final.pdf, accessed June 18, 2012.

31. Watson PJ, Brymer MJ, Bonanno GA. Postdisaster psychological intervention since 9/11. Am Psychol 2011;66(6):482–494.

32. National Commission on Terrorist Attacks upon the United States. The 9/11 Commission Report. Zelikow P, Jenkins BD, May ER, editors. Washington, DC: 2004. p 285.

33. Averill JD, Mileti D, Peacock R, et al. Federal investigation of the evacuation of the World Trade Center on September 11, 2001, 3rd International Conference Proceedings. In: Waldau N, Gattermann P, Knoflacher H, Schreckenberg M, editors. *Pedestrian and Evacuation Dynamics*. New York: Springer; 2007. pp 1–12.

34. New York City Department of Health and Mental Hygiene, Office of Chief Medical Examiner. Update on the results of DNA testing of remains recovered at the World Trade Center site and surrounding area, February 1, 2009. Available at http://www.nyc.gov/html/ocme/downloads/pdf/public_affairs_ocme_pr_february_2009.pdf, accessed June 18, 2012.

35. 9/11 by the numbers: Death, destruction, charity, salvation, war, money, real estate, spouses, babies, and other September 11 statistics. New York Magazine, 2002. Available at http://nymag.com/news/articles/wtc/1year/numbers.htm, accessed June 18, 2012.

36. Ibid.

37. Neria Y, Gross R, Litz B, et al. Prevalence and psychological correlates of complicated grief among bereaved adults 2.5–3.5 years after September 11th attacks. J Trauma Stress 2007;20(3):251–262.

38. Schlenger WE, Jordan BK, Caddell JM, et al. Epidemiological methods for assessing trauma and PTSD. In: Wilson JP, Keane TM, editors. *Assessing Psychological Trauma and PTSD*. 2nd ed. New York: Guilford Press; 2004. pp 226–261.

39. Silver RC, Holman EA, McIntosh DN, et al. Nationwide longitudinal study of psychological responses to September 11. JAMA 2002;288(10):1235–1244.

40. 9/11 by the numbers.

41. Neria.

42. Centers for Disease Control and Prevention. Rapid assessment of injuries among survivors of the terrorist attack on the World Trade Center—New York City, September 2001. MMWR Weekly, 2002;51(1):1–5.

43. Ibid.

44. Moline J, Herbert R, Nguyen N. Health consequences of the September 11 World Trade Center attacks: A review. Cancer Invest. 2006;24:294–301.

45. Bram J, Orr J, Rapaport C. Measuring the effects of the September 11 attack on New York City. Econ Pol Rev 2002;8(2):5–20.

46. Ibid.

47. Simonsen LF, Reyes G. *Community-Based Psychological Support: A Training Manual*. Geneva, Switzerland: International Federation of Red Cross and Red Crescent Societies; 2003.

48. Stith A, Panzer AM, Goldfrank LR, editors. *Preparing for the Psychological Consequences of Terrorism: A Public Health Strategy. Committee on Responding to the Psychological Consequences of Terrorism, Institute of Medicine.* Washington, DC: National Academies Press; 2003.

49. Disaster Mental Health Subcommittee of the National Biodefense Science Board, 2010. Integrating behavioral health in federal disaster preparedness, response, and recovery: Assessment and recommendations. Available at http://www.phe.gov/preparedness/legal/boards/nbsb/meetings/documents/dmhreport1010.pdf, accessed June 18, 2012.

50. Elrod CL, Hamblen JL, Norris FH. Challenges in implementing disaster mental health programs: State program directors' perspectives. Ann Am Acad Pol Soc Sci 2006;604:152–170.

51. Ruscher JB. Stranded by Katrina: Past and present. Anal Soc Issu Pub Pol 2006;6(1):1–6.

52. Stuber J, Galea S, Boscarino JA, Schlesinger M. Was there unmet mental health need after the September 11, 2001 terrorist attacks? Soc Psychiatry Psychiatr Epidemiol 2006;41(3):230–240.

53. Frank RG, Pindyck T, Donahue SA, et al. Impact of a media campaign for disaster mental health counseling in post-September 11 New York. Psychiatr Serv 2006; 57(9):1304–1308.

PART III

INSTITUTIONAL MANAGEMENT

9

THE ROLE OF THE MANAGER OF MASS CASUALTY AND DISASTER EVENTS

Shmuel C. Shapira and Limor Aharonson-Daniel

THE WEDDING HALL COLLAPSE

On Thursday, May 24, 2001, at 10:43 P.M., the third floor of the Versailles Wedding Hall in Jerusalem collapsed. Midway through a wedding celebration, many of the dancing guests plunged through a massive hole in the floor, although the walls and supporting columns of the building remained intact. Among the 700 people present, 23 died and 315 were injured. The bride suffered a fractured pelvis.

The initial impression of the first responders was that the event was related to a terror attack. Terror assaults were very common in Israel during that period, especially in the capital, Jerusalem. At first, emergency medical technicians (EMTs) believed that a huge bomb might have hit the building. But an investigation soon after suggested otherwise: no explosion had been heard and no trace explosives were detected.

Eventually the cause was determined to be faulty construction. The testimony of survivors and videos of the event revealed that several victims fell gradually through the second floor on to the first. At about 11:30 P.M., one of the authors (Shapira) received a phone call from an anxious friend, a woman who had attended the wedding. Although she had fallen through two floors she was only slightly injured. But she was looking for her 12-year-old son, who had also been at the wedding. Four hours later he was located at one of Jerusalem's four hospitals. He was more seriously wounded than his mother and was later transferred to another of the hospitals, a level I trauma center with more advanced capabilities to treat complicated injuries.

The transfer exemplifies principles that Israel has developed through experience with numerous responses to terrorist and disaster events. Victims at an event site are quickly triaged to a particular hospital based on the perceived severity of injury, but also with an

Local Planning for Terror and Disaster: From Bioterrorism to Earthquakes, First Edition.
Edited by Leonard A. Cole and Nancy D. Connell.
© 2012 John Wiley & Sons, Inc. Published 2012 by John Wiley & Sons, Inc.

effort to avoid overloading any single hospital. If necessary as a corrective, secondary triage to another hospital may later be undertaken.

The first victim of the Versailles building collapse arrived at a hospital at 11:06 P.M. Within the first hour, 130 (42%) of the casualties had been evacuated from the collapse site, and during the second hour another 105 (33%) were evacuated. Patients continued to arrive at the hospitals until 3:00 A.M., more than 4 hours after the collapse. (In subsequent years, after much more experience with terror attacks, evacuation times from incident sites were dramatically shortened.) Thirty-two victims were subject to secondary triage—movement from one of the Jerusalem hospitals to another. The emergency medical service (EMS) regional center dispatched 121 ambulances and 655 staff members. The hospital personnel on duty were supplemented by about 1300 extra medical and paramedical staff.[1]

In May 2007, four engineers involved in the construction failures at the Versailles Wedding Hall were found guilty of negligence and sentenced variously to between 2 and 4 years of prison.

Several questions arise that are relevant to management of such an event:

- Who should manage the prehospital operation of a mass casualty event (MCE)?
- Who should manage the hospital MCE operation?
- How do managers deal with an MCE?
- What are the professional networks involved in MCE management?
- What is the role of primary triage?
- When and how is a decision on secondary triage made?
- What is the concept of a "triaging hospital"?
- What are the potential pitfalls during the management of an MCE?
- How to prepare for an MCE and for sudden disasters?

BACKGROUND

Decades of experience with terror attacks in Israel have generated cumulative knowledge concerning preparations for terror and non-terror related MCEs.[2,3] This knowledge has been abetted by Israeli experience with full-scale military conflict and rescue expeditions to disaster zones in other countries.[4,5] These engagements involved ongoing risk assessments, quick decision-making, and the development of operational skills for handling MCEs. Altogether they have produced a valuable platform for crisis management and leadership.[6,7]

The mission of the medical leader under such circumstances is to optimize the system in order to save lives, decrease morbidity, reduce permanent disability, and resume routine medical system function as quickly as possible. Optimal function of a medical system under stressful circumstances can enhance the resilience of a society and enable it to better withstand hardships associated with terrorism and other forms of disaster.[8] The tedious process mandates absolute commitment of managers at the different national and institutional levels to prepare for dire events.

The Optimal Manager for MCEs

A primary concern with an MCE or disaster is the resulting number of casualties. Thus, a key to managing an event lies with the availability of necessary resources, a condition made more tenuous when an event is unexpected. Resources can be classified as human (cognitive, knowledge-based, and emotional) and tangible (facilities, equipment, and supplies). Establishing or restoring a balance between needs and resources can depend on the wisdom, experience, and communication skills of the manager. In order to fulfill these roles, a manager should possess specific attributes:

- Ability to recognize relevant threats (man-made and natural) and their implications
- Familiarity with daily operational routines and manpower
- Familiarity with standard operating procedures (SOPs)
- Experience with management of previous events, or at least mock events during drills

Ideally, the manager should be selected by those at the top of the managerial pyramid, such as the hospital director and head of the EMS region, in advance of any event. At least one additional senior functionary should be named to act in the manager's absence. These "shadow" managers should have the training and capability to independently manage in an MCE and disaster situation. If circumstances oblige, the shadow manager initially should take command; when the designated principal manager arrives on-scene, he may or may not assume responsibility. In general, the default decision should be for the arriving manager to assist the more junior peer and not take over in the midst of an event. In Israel, the typical choice for field commander of an MCE is someone at the level of police chief inspector or regional chief officer. During an unusually intense event, such as an earthquake, the head of the Home Front Command or his deputy may take charge.

Responsibility for the overall medical aspects of an event lies with the prehospital medical commander: the EMS regional director or his deputy. Under exceptional circumstances, this responsibility might be assumed by the EMS director general or the EMS director for operations. Normally, the first senior medical official to arrive on-scene becomes the sole commander of the medical teams at that site. This official announces himself on the radio network as "command 10," thus informing all responder groups of his identity. Furthermore, the official dons a distinctive hat so that his position is obvious to everyone at the site.

A frontal command post should be established at the site and staffed by representatives of all the responder groups including the fire brigade, EMS, and officials of the municipal government. At the hospital, MCE management is led by the hospital director or deputy director. It is important to emphasize that whether at the event site or in the hospital, no director can effectively operate alone. Successful management depends on teams of skilled and dedicated personnel.

Leadership

Hospital and EMS organizational structures are hierarchical with defined chains of command and responsibility. But leadership emanating from the top will influence effectiveness throughout the organization. Marcus et al. review the commonly perceived direction of leadership in a crisis situation. The complexities of leadership are often obscured by an inclination to view the process simply as top-down, with leaders leading followers.[9] According to this understanding, the boss-to-employee relationship is formalized in clear roles, rules, job descriptions, and responsibilities that prescribe performance. But in fact, leaders often act in ways that transcend usual organizational confines.[10]

This is largely because for the sake of improved productivity there is often need for participation by people who work in different sectors or different levels of a structured hierarchical framework. Linking the efforts of people who may be from disconnected organizational units, leverages their activities into something that would not otherwise be achievable.[11] The capacity to attain these linkages beyond the confines of their own bureaucratic entities is a characteristic of good leadership. Identification of the gaps, between what should be done and the will to do it, can be instrumental in achieving an otherwise unachievable result.[12]

Teamwork is important in disaster response and one aim of a leader is to encourage it during preparedness exercises. Teamwork is especially critical in the rendering of medical care and can make the difference between life and death in emergency situations. A recent study suggests that when individuals identify strongly with a group, their contributions may compensate for lapses by others in the group.[13]

Key Responsibilities of the EMS Field Manager

Upon arrival at an event site, the designated manager assumes command and immediately provides information to EMS headquarters about the circumstances on the ground. The initial report should include a description of what happened—whether an explosion, shootings, a train accident. The report should indicate the precise location, estimated numbers and severity of casualties, and other relevant information such as whether shooting (if that was the cause) is ongoing and the site is accessible.

Following this initial report to headquarters, the EMS manager should provide his immediate subordinates with initial instructions including policies for safety, assignment of medical staff to specific locations, and preliminary guidelines for treatment (decontamination, provision of antidotes, etc.). This initial phase should last only a few minutes. Next, the EMS manager should assess needs and available resources and, if necessary, request more aid from the regional and national headquarters.

Needs are largely a function of the number and types of injuries. Resources to match the needs include available medical teams, equipment, ambulances, and helicopters. At this stage the most important concern for the local manager is performance of primary triage. Triage in the field aims at sending a victim to an appropriate medical facility in accordance with the nature and severity of one's injury.[14] At the same time patients should be distributed among several hospitals to avoid overloading any one facility.[15,16] Thus the EMS manager must remain informed in real time about the capabilities of both nearby and more distant hospitals. This information becomes central to decisions about patient distribution.

Primary triage, a core initial task of the EMS, should be the sole responsibility of a designated senior emergency physician or paramedic. The field manager himself should not perform triage since this could preclude his attending to other responsibilities. Israeli experience has shown the wisdom of simplifying triage to only two categories: urgent and non-urgent.[17] These designations dictate the order of evacuation. The modus operandi for treatment is "scoop and run"—providing minimal intervention only for the most urgent cases and then rapid evacuation.[18] Most patients are evacuated by ambulance, though some might be transported in private cars.

During disasters, when a site is remote or accessibility is difficult, helicopters might be the best or only available means of evacuation.[19] When available, they can move beyond traffic jams, a common challenge during an MCE. Air-medical evacuation also enables movement of casualties to more remote trauma centers, which also helps avoid overcrowding at nearby hospitals. Ultimately, the triage officer determines each ambulance's destination.

It is essential that EMS managers have the means to communicate with relevant individuals. In Israel, communication networks are capable of connecting EMS personnel with hospitals, the Home Front Command, police, and municipal information and resilience centers. Moreover, all emergency organizations are connected through a single paging system. Thus, notification of an event is sent simultaneously to all concerned parties.

Although Israeli communication networks are continuously available, some are not operated routinely. But these networks are checked frequently to ensure functionality in case of an emergency. Through ongoing communication with the police, fire brigade, and HAZMAT (hazardous materials) units, the commander can update subordinates about safety measures and evacuation routes. In addition to the radio channel serving EMS and medical efforts, an additional secure radio network serves all Israeli emergency first responders including the army.[20]

Key Responsibilities of the Hospital Manager during an MCE

During an MCE, the hospital manager must oversee rapid transformation of a slow and complex system to one that addresses a sudden influx of patients in urgent need of care. Meeting the challenge effectively is possible only if the hospital staff and infrastructure are well prepared for that moment. Early information available to the hospital director is commonly equivocal and incomplete. But following an alert that an event is imminent or underway, the response process begins.

Hospital operations during an MCE are optimally managed through two command posts: a front station near the entrance to the emergency department and a back station in one of the hospital boardrooms. At the front post, the hospital director or his deputy, assisted by a few associates, can directly view and help control the patient influx. Donned in vests that identify their commander status, these individuals are equipped with portable radios and loudspeakers. The goal is to facilitate the admission process and activate special measures such as decontamination or patient isolation, as needed.

The back station is in charge of hospital communications both within the facility and with outside parties. This station is headed by the hospital director or his deputy and staffed by senior managers representing all hospital divisions. Following the screening of patients, one of the hospital director's prime efforts is to anticipate and avoid the formation of bottlenecks.

In addition to medically related activities, the hospital MCE manager must oversee a variety of other tasks including

- *Hospital Security and Public Order*. It is crucial to maintain safety and sound operation while respecting the privacy rights of patients.
- *Establishment of a Public Information Center*. The center provides information by phone or in person to worried families and other outside inquirers. It should be able to accommodate a large number of people and, to avoid congestion, be located at a distance from the emergency department.[21]
- *Periodic Dissemination of Press Releases and Relevant Statements*. In conjunction with the information center, the manager facilitates the availability of hospital spokespersons for media interviews and updates.

ADMISSION AREAS, SECONDARY TRIAGE, THE "TRIAGE HOSPITAL"

A key challenge for a hospital director at the outset of an MCE is whether to open additional areas for admission to the hospital. The patient influx can be so extensive that supplementary locations become necessary. Optional sites for admission (commonly hospital lobbies) should be part of an institution's preparedness plan. But while additional admission areas accelerate initial processing, they also complicate management. That is because staff members function most effectively in a customary work environment rather than an improvised one. Thus, additional admission sites should be set up only when initial processing has been markedly slowed by an unusually large numbers of incoming patients.

After arrival at a hospital, patients may experience a secondary triage, which could mean dispatching them to another hospital. The purpose is to decrease the patient load in the primary admitting facility or to send someone to a hospital better able to offer specialized care.[22] Secondary triage may also be implemented to correct mistakes made during primary triage.[23] In any case, a decision to employ secondary triage should be based entirely on the welfare of the patient and not on other ostensible institutional interests.[24]

The concept of a "triage hospital" originates from the field of military medicine. The triaging hospital functions as a big emergency department. Only patients requiring life-saving interventions are admitted, evaluated, and treated. A victim who is transportable and not suitable for release in less than an hour after arrival will be transferred to another hospital. The decision to operate a hospital in the mode of triaging is a major determination that should be supported by the local health authorities. Once the determination is made, it should be communicated immediately to the EMS and neighboring hospitals. In a large-scale event, the default designation of a triage hospital should be one near the event site. A level I trauma center should never be designated a triaging hospital, since this would preclude patients' benefiting from its advanced medical capabilities.

POTENTIAL PITFALLS DURING MCE MANAGEMENT

Distraction

Once SOPs are in place, the manager must focus on assuring that they are being performed. Anything that distracts from overall management during the MCE could hamper the outcome. Thus, for example, the manager should neither provide direct care nor participate in decision-making related to the management of an individual victim.

Absence of a Threshold for Avoiding Treatment

The mission in an MCE is to save as many lives as possible. At times, this may mean sacrificing certain critically injured individuals who might receive treatment under other circumstances. A decision not to provide treatment other than pain management is based on the severity of injury, probability of survival, or the resources that would be diverted from treating others with greater chances of survival. In a radiological event, for example, patients exposed to 8–10 grays* or more would receive only palliative care.[25] That is because no matter what manner of care whole body exposure to such radiation levels means almost certain death in a matter of weeks.

---

*The gray is a unit of absorbed radiation dose of ionizing radiation (e.g., X-rays) and is defined as the absorption of one joule of ionizing radiation by one kilogram of human tissue.

Still, the event manager should consult with the most senior on-site physician and with regional health authorities, before making a decision about a cut-off level for treatment. After consultation, the manager can inform the medical teams about the designated threshold for comfort care only. Triage thresholds might also be applicable to prehospital care when caregivers and life support equipment are in short supply. Scarcity of ventilators and other life-saving apparatus would of course affect ability to provide optimal treatment.

A choice to downgrade from optimal to lesser care even for a limited period is in part a moral consideration. Thus, it is all the more important to establish proposed courses of action for dire situations in the planning stage, before facing the actual event. The moral component adds to the difficult challenge of decision-making. Experienced, knowledgeable, and sensitive medical leadership is essential to the task.

Absence of Flexibility and an Open Mind Despite Pressure

Contingency plans can enable a hospital to function beyond its routine capabilities. They must include allowance for utilization of additional personnel (largely off-duty staff), extra space (lobbies, corridors, dining rooms), and obtaining additional equipment and medications from local and national stockpiles. Gaps and errors in the process can best be identified through exercises, so that rectifications are possible in advance of a real event.

In the case of an EMS, formation of extra capacity could mean evacuation of more than one victim in an ambulance, or transporting a group of those with minor wounds in a bus, with minimal medical escort.

There is a limit to the ability of a hospital and EMS to expand capacity. At some point, treatment and evacuation times will be suboptimal and may reach an unacceptable level. The MCE manager should take early measures to avoid, or at least delay, reaching this point, such as declaring the facility to be a triaging hospital.

A key quality in management of complex situations is flexibility of mindset. This could mean not only departure from a daily routine, but in some circumstances from the MCE plan itself.

Bureaucracies and Bottlenecks

SOPs for an MCE should be kept simple and as similar as possible to daily operational routines. Bureaucratic obstacles should be bypassed. Some level of disorder is inevitable during a sudden and complex situation, but avoiding bottlenecks helps to minimize chaos. The EMS manager must be wary of inheriting the chaos of the prehospital scene. An important potential bottleneck is a shortage of ambulances. One approach to mitigating this problem could be to call for additional ambulances from neighboring EMS regions. Another would be to send more patients to the nearest hospital even if the result is slight overloading. This would allow ambulances more quickly to return to the scene for additional victims.

Ongoing communications with the police provides information about traffic concentrations and preferred evacuation routes. A distinctive EMS radio channel is also available for use by the MCE manager. At the same time, the EMS could carry out its routine activities and communicate via its regular operative networks.[26] In fact, daily medical priorities should not be degraded when weighed against MCE needs. A patient with acute myocardial infarction is no less in need of urgent care and evacuation than a seriously injured terror victim.

In-hospital bottlenecks are most likely to occur in the admission areas. Other possible sites include the operating room, imaging area, and intensive care unit.[27,28] The manager should have real-time information on patient flow and resource utilization. Israeli experience shows that bottlenecks are less likely with the implementation of unidirectional patient flow. Thus, a patient who is sent from the emergency department for a computerized tomography (CT) scan would not return to the emergency department. Rather, a senior surgeon stationed at the imaging department would determine the next destination for the patient, whether to an operating room, the intensive care unit, or perhaps to the ward.

When a Hospital Is Directly Affected by an Event

A hospital might be directly affected by an earthquake, a missile, or a cloud of toxic chemicals. Under these circumstances, the hospital's ability to provide medical services could be sharply curtailed. Services to the surrounding community might be suspended entirely, while the staff assists already-hospitalized patients, visitors, and others already on the premises. Here too, the event manager must provide leadership within the limits of available resources.

A memorable challenge involved Dr. Anna Pou at the Memorial Medical Center in New Orleans during Hurricane Katrina in August 2005. Electric power was lost, which deprived the hospital of air conditioning during a period of sweltering heat. Refrigerators grew warm and stored items including blood products and medications could not be preserved. An overflowing sewage system caused a trenchant odor throughout the building.[29,30] In the absence of electricity, ventilators ceased to function, which was especially dangerous for patients in intensive care, a unit under Dr. Pou's management.

After several days, the staff decided to evacuate the hospital, where several patients had already died. Movement of some of the patients on upper floors was slow and difficult. Afterward, allegations were raised that some of the patients had deliberately been given fatal doses of morphine and other drugs. Eventually, a grand jury heard witnesses, reviewed the evidence, and declined to indict Dr. Pou or any of the other suspects.

As a manager, Dr. Pou had to think of the patients, with no idea if and when help would come. She had to think of the staff, trying to provide medical care under desperate conditions: hot, hungry, thirsty, lacking resources. Given the high mortality rate, she then had to face the accusation of euthanasia. Such scenarios should be studied and considered during the planning for management of a disaster situation.

In her own response to the event, Dr. Pou urged that there be more focus on the training of civilian physicians in disaster triage, education regarding military evacuation protocols, and plans that are tested and then actually followed during a crisis. Not surprisingly, she also called for clearer medical and ethical guidelines for disaster care. And she urged greater protection for doctors and nurses from legal suits related to service during a federally declared emergency.[30]

PREPARATION

Knowledge and Risk Management

Management of MCEs and disasters are not part of the usual administrative or medical curriculum. The subject, therefore, must be presented in a special program that assumes no previous knowledge of the area.

An introductory discussion should describe threats, risk analysis, and preparation regarding these events.[31] Threats variously include

- terrorism and armed conflict that might involve conventional explosives, or biological, chemical, radiological, or cyber mechanisms
- man-caused or accidental large-scale transportation damage
- man-caused or accidental large-scale industrial damage
- natural disasters including earthquakes, fire, tsunamis, storms, floods, and heat waves

In conducting risk analysis, risk is thought of as the probability of the event occurring multiplied by its anticipated consequences (the cost, in lives or dollars).

Establishing SOPs and Checklists

After defining a threat, response organizations should develop their own written SOPs. Some procedures could be the same for several agencies, such as a notification system to secure additional workers if staff reinforcement becomes necessary. But other parts of the SOPs would apply only to certain scenarios. Thus chemical, radiological, or biological exposures would each be uniquely addressed. Furthermore, while an SOP should cover operations by individual medical and administrative departments, it should also address the manner that these departments would be communicating and coordinating with each other.[32]

The SOP should be updated every year or two, or sooner if circumstances warrant. This could be the case with a biological or chemical agent that has been newly identified as a threat, or with a hospital that has undergone major structural changes.

A written checklist, while a simple adjunct of an SOP, is imperative. Especially given the stressful atmosphere of an MCE, a manager cannot rely on memory alone to address every procedural requirement. Thus, a checklist should be in the hands of managers at every event location.

Ongoing Education and Training

After the modus operandi for dealing with an event is clear to the leadership, it should be widely circulated. Education and training is essential both within disciplines and across disciplinary lines. Instruction within a field focuses on information specific to that field—training for physicians, nurses, and other hospital staff would concentrate on each of their discipline's role in an MCE. But training must also include interdisciplinary planning and activity. Responders must be able to use a common language and have an understanding of the roles of fellow workers. As noted, effective communication is the key to successful management of MCEs and disasters.

Training, therefore, should include both types of workshops: one that is narrowly framed for participants with a common professional background, and the other more general for responders from a variety of fields. In both cases, training can benefit from the use of multimedia resources. Software is available to enhance interest, create modeling tools, and ultimately to evaluate knowledge and understanding.[33–35]

Exercises and Drills

The principal instrument for the development of hands-on capability and organizational preparedness is the drill.[36] A table-top exercise is a common and inexpensive form that while limited, can be effective.[37] Participants gather around a table and react to a sequence of challenges posed by a moderator. For management training, the drill enables 15–20 relevant managers to make key decisions as an MCE scenario unfolds. The moderator may narrow both the theoretical resources available to the participants and the time permitted to make decisions. By imposing pressure on the decision-makers, the drill can induce a stressful atmosphere reflective of a real event.

For general system training, the best form of drill is an action exercise that includes participation from numerous responder fields. Physicians, paramedics, nurses, and other responders may engage with simulated casualties.[38] Mock victims can be passive or, in more realistic (and expensive) scenarios, medically educated actors. These "smart" victims present symptoms and then react to the care provided to them. Their behavior provides valuable information not only about medical treatment but also about equipment shortages and other possible obstacles that surface during the exercise.[39] Among the total number of simulated casualties, even if only 20% are smart victims, they add a measure of realism to a full-scale exercise.

Debriefing

A debriefing meeting should follow every simulated or real MCE. The aim is to draw lessens from the experience, match them against existing protocols, and make changes, if needed, to the SOPs. The debriefing should be carried out as soon as possible after the event: the earlier it takes place, the fresher the memory of the participants. Moreover, improvements can then be instituted all the more quickly.

All relevant MCE managers should participate in the debriefing process. Discussion should be open and uninhibited. If individuals offer disparate interpretations of certain activities, efforts should be made to overcome the discrepancies and arrive at a consensus position. Even if not possible to resolve the differences, the value of open discussion remains undiminished.

In recent years, concerns have surfaced about the psychological effects on participants in debriefings who themselves may have been traumatized by a real event. Whether debriefing has helped or hindered their ability to process their own emotions remains a matter of dispute. The uncertainty has led some groups to suspend the practice of debriefing, though most organizations have not. The value of debriefings has been affirmed by military, emergency services, and humanitarian aid organizations.[40] Before adopting a policy to end the procedure, a careful assessment should weigh its purported negative effects on some individuals against its demonstrated value.

CONCLUSION

Mass casualty disaster events often occur unexpectedly and thus require ongoing readiness. Successful management relies not only on actions taken during and after an event but also on the steps taken before one occurs. Planning for various scenarios, anticipating a range of potential developments, and engaging in exercises are central to successful outcomes. Still, no matter how talented the manager, suitable equipment and supplies along with trained support staff are also essential.

In order to achieve optimal response, it is important for each organization to have updated SOPs. These help in thinking through the planning process, assessing proficiency during drills, and directing the response effort in a real event. SOPs are necessary though not sufficient for optimal management of a mass casualty or disaster event. Human attributes are also essential including respected leadership, a sense of professionalism, team spirit, and dedication to duty.

NOTES

1. Avitzour M, Libergal M, Assaf J, et al. A multicasualty event: Out-of-hospital and in-hospital organizational aspects. Acad Emerg Med 2004;11:1102–1104.

2. Shapira SC, Hammond JS, Cole LA. *Essentials of Terror Medicine*. New York: Springer; 2009.

3. Falk O, Morgenstern H. *Suicide Terrorism*. New York: Wiley; 2009.

4. Leiba A, Ashkenasi I, Nakash G, et al. Response of Thai hospitals to the tsunami disaster. Prehosp Disaster Med 2006;21(1):32–37.

5. Ginosar Y, Shapira SC. The role of an anaesthetist in a field hospital during the cholera epidemic among Rwandan refugees in Goma. Br J Anaesthesia 1995;75:810–816.

6. Shapira SC, Mor-Yosef S. Terror politics and medicine—The role of leadership. Stud Confl Terrorism 2003;27:65–71.

7. Schreiber S, Yoeli N, Paz G, et al. Hospital preparedness for possible nonconventional casualties: An Israeli experience. Gen Hosp Psychiatry 2004;26(5):359–366.

8. Norris FH, Stevens SP, Pfefferbaum B, et al. Community resilience as a metaphor, theory, set of capacities, and strategy for disaster readiness. Am J Community Psychol 2008;41(1–2): 127–150.

9. Fernandez RM. Structural bases of leadership in intraorganizational networks. Soc Psychol Q 1991;54(1):36–53.

10. Marcus LJ, Dorn BC, Henderson JM. Meta-leadership and national emergency preparedness: A model to build government connectivity. Biosecur Bioterror 2006;4(2):128–134.

11. Schein EH. *Organizational Culture and Leadership*. San Francisco, CA: Jossey-Bass; 2004.

12. Kotter J. *Leading Change*. Boston, MA: Harvard Business School Press; 1996.

13. Fishbach A, Henderson MD, Koo M. Pursuing goals with others: Group identification and motivation resulting from things done versus things left undone. J Exp Psychol 2011;140(3): 520–534.

14. Mackersie RC. History of trauma field triage development and the American College of Surgeons Criteria. Prehosp Emerg Care 2006;10(3):287–294.

15. Cole J. Medical response to terrorism. In: Richman A, Shapira SC, Sharan Y, editors. *Medical Response to Terror Threats*. Amsterdam: IOS Press; 2010.

16. Einav S, Feigenberg Z, Weissman C, et al. Evacuation priorities in mass casualty terror-related events: Implications for contingency planning. Ann Surg 2004;239(3):304–310.

17. Raiter Y, Farfel A, Lehavi O, et al. Mass casualty incident management, triage, injury distribution of casualties and rate of arrival of casualties at the hospitals: Lessons from a suicide bomber attack in downtown Tel Aviv. Emerg Med J 2008;25(4):225–229.

18. Gold CR. Prehospital advanced life support vs "scoop and run" in trauma management. Ann Emerg Med 1987;16(7):797–801.

19. Assa A, Landau DA, Barenboim E, Goldstein L. Role of air-medical evacuation in mass-casualty incidents—A train collision experience. Prehosp Disaster Med 2009;24(3):271–276.

20. Report of the 7 July Review Committee, London Assembly, May 2006.

21. Lynn M, Gurr D, Memon A, Kaliff J. Management of conventional mass casualty incidents: Ten commandments for hospital planning. J Burn Care Res 2006;27(5):649–658.

22. Mekel M, Bumenfeld A, Feigenberg Z, et al. Terrorist suicide bombings: Lessons learned in Metropolitan Haifa from September 2000 to January 2006. Am J Disaster Med 2009; 4(4):233–248.

23. Schwartz D, Pinkert M, Leiba A, et al. Significance of a Level-2, "selective, secondary evacuation" hospital during a peripheral town terrorist attack. Prehosp Disaster Med 2007;22(1):59–66.

24. Avitzour M, Aharonson-Daniel L, Peleg K. Secondary transfer of trauma patients: Rationale and characteristics. Isr Med Assoc J 2006;8:539–542.

25. Mettler FA, Guskova AK. Treatment of acute radiation syndrome. In: Gusev IA, Guskova AK, Mettler FA, editors. *Medical Management of Radiation Accidents*. 2nd ed. Washington, DC: CRC Press; 2001.

26. Shapira SC, Shemer J. Medical management of terrorist attacks. Israel Med Assoc J 2002; 4:489–492.

27. Hirshberg A, Holcomb JB, Mattox KL. Hospital care in multi-casualty incidents: A critical view. Ann Emerg Med 2001;37:647–652.

28. Kosashvili Y, Aharonson-Daniel L, Peleg K, et al. Israeli hospital preparedness for terrorism-related multiple casualty incidents: Can the surge capacity and injury severity distribution be better predicted? Injury 2009;40(7):727–731.

29. Okie S. Dr. Pou and the hurricane—Implications for patient care during disasters. N Engl J Med 2008;358:1–5

30. Pou AM. Hurricane Katrina and disaster preparedness. N Engl J Med 2008;358:1524.

31. Morgan MG. Risk analysis and management. Sci Am 1993;269(1):24–30.

32. Adini B, Goldberg A, Laor D, et al. Factors that may influence the preparation of standards of procedures for dealing with mass-casualty incidents. Prehosp Disaster Med 2007;22(3): 175–180.

33. Knight JF, Carley S, Tregunna B, et al. Serious gaming technology in major incident triage training: A pragmatic controlled trial. Resuscitation 2010;81(9):1175–1179.

34. Heinrichs WL, Youngblood P, Harter P, et al. Training healthcare personnel for mass-casualty incidents in a virtual emergency department: VED II. Prehosp Disaster Med 2010;25(5):424–432

35. Hannon B, Ruth M. *Dynamic Modeling*. New York: Springer; 2001.

36. Kaji AH, Lanford V, Lewis RJ. Assessing hospital disaster preparedness: A comparison of an on-site survey, directly observed drill performance, and video analysis of teamwork. Ann Emerg Med 2008;52:195–201.

37. O'Neill PA. The ABC's of disaster response. Scand J Surg 2005;94(4):259–266.

38. American College of Surgeons. *Advanced Trauma Life Support*. 8th ed. Chicago, IL: American College of Surgeons; 2008.

39. Gofrit ON, Leibovici D, Shemer J, et al. The efficacy of integrating "smart simulated casualties" in hospital disaster drills. Prehosp Disaster Med 1997;12(2):97–101.

40. Hawker DM, Durkin J, Hawker DS. To debrief or not to debrief our heroes: That is the question. Clin Psychol Psychother 2010;18:453–463.

10

THE ROLE OF PUBLIC HEALTH

Henry Falk and Isaac Ashkenazi

Public health has very broad responsibilities in preparedness for and response to terrorism and disasters. Many of these responsibilities are common to all terror/disaster scenarios and fall under the rubric of an all-hazards approach. Responses to the Chernobyl nuclear plant accident, Hurricane Katrina, 9/11, and the 2001 anthrax attacks all share features. At the same time, many actions required skills that were geared to the individual event and its context. This chapter surveys the public health response to the Madrid train bombings of 2004, followed by discussion of other terrorist explosive events and of disasters arising from natural or accidental causes. The concluding portions synthesize the various public health actions in these scenarios to provide a framework for the public health role within the broader context of a government-wide and societal response to such disasters.

THE MADRID TRAIN BOMBINGS

On March 11, 2004, ten explosive devices were detonated in rapid succession in four commuter trains in or adjacent to four Madrid train stations. The bombings killed 191 people. Among them, 177 died on scene and the remainder succumbed during transit to or in hospital while under treatment.[1] More than 2000 people were injured, with more than 300 on scene considered critically or seriously wounded; hundreds were taken to emergency departments or briefly hospitalized for limited injuries; almost 100 patients were hospitalized for further treatment of extensive injuries.[2]

Figure 10.1 depicts the distribution and timing of bomb attacks among the stations. More than 200 ambulances and emergency vehicles converged on the scene to transport patients to a hospital. Several times as many emergency management services (EMS) workers provided on-scene treatment and triage. Figure 10.2 highlights the intensity of this EMS response at one of the four stations involved (Tellez), where 65 died and 165 (55 critically and 110 seriously injured) received on-site medical assistance. All the patients from this site were

Local Planning for Terror and Disaster: From Bioterrorism to Earthquakes, First Edition.
Edited by Leonard A. Cole and Nancy D. Connell.
© 2012 John Wiley & Sons, Inc. Published 2012 by John Wiley & Sons, Inc.

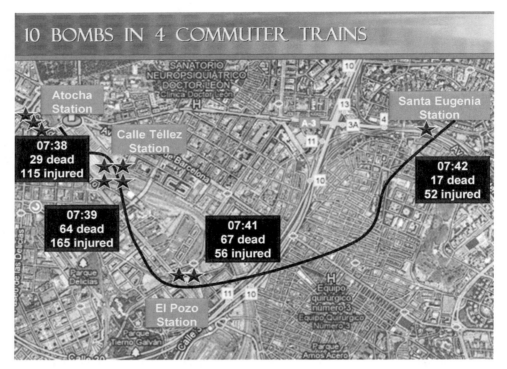

FIGURE 10.1   Depiction of distribution and timing of bomb blasts at four Madrid train stations and number of fatalities resulting from the blasts. (Courtesy of Dr. Fernando Turegano Fuentes.) (For a color version of this figure, see the color plate section.)

then transported to the Gregorio Maranon Hospital. This was a particularly severe blast, as the train was in motion with the doors closed at the time of the explosion, magnifying the blast injury. Figure 10.3 depicts the field triage site where patients were grouped adjacent to the train and then brought to a nearby sports complex building that served as an advance medical post.

Health issues that were central to this event—on-site emergency care, over triage and under triage, "scoop and run" versus "stay and play," distribution and transportation of casualties to the available medical facilities, in-hospital treatment, attention to the worried well and family members, and respectful disposition of the dead—are common to many terrorist bombings and are therefore key aspects of public health preparedness planning. The assurance of appropriate and rapid responses linked to non-health response groups is also central to preparedness planning.

EMS and hospital activities have been variously recounted and the experience at the Gregorio Maranon University Hospital has been described in particular.[3] These articles note the fortunate timing of the event insofar as it occurred during a change of shift. Thus, more personnel than usual were available to address challenges including operating room capacities and other matters of patient care. They also addressed the activation of the hospital disaster plan, though bottlenecks nevertheless arose from the surge in demand for care.

The management and care of critically injured patients in terrorist bombings includes addressing many expected sources of injuries (flying debris, shrapnel, bomb fragments,

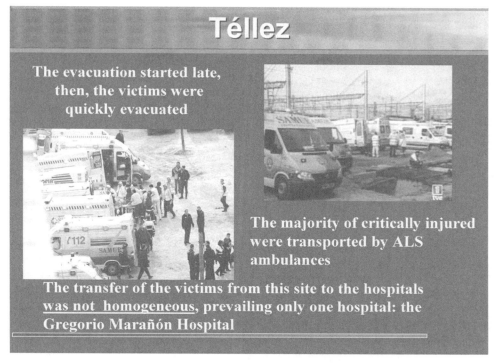

FIGURE 10.2   Photos of EMS response on scene at Tellez train station site. (Courtesy of Dr. Ervigio Corral Torres; Presented at a Tale of Cities meeting, New York, March 15, 2011. Available at http://www.taleofourcities-newyork.com/presentations/Day1/Torres%20NY%20March%202011.pdf.) (For a color version of this figure, see the color plate section.)

falls, burns) as well as severe but often initially unperceived wounds. These include blast lung and abdominal injuries and rupture of the tympanic membranes, which impair a victim's hearing and can impede communications with emergency care workers.[4] The surge in demand for vital services was exemplified by the 37 major surgical interventions on 34 victims performed in a single Madrid hospital.[5]

In a presentation at a "Tale of Cities" meeting it was noted that during the first 24 hours after the bombings, in the seven receiving hospitals, 82 victims underwent 124 major surgical interventions—most for orthopedic/trauma, plastic, abdominal, maxillofacial, neurosurgical, or ophthalmic injuries. (The Tale of Cities project brings together public health and other emergency response leaders from several countries with deep experience in responding to terrorist explosive events, to dialogue and share experiences.[6])

EMS issues are also addressed in three articles that take very different views of the efficacy of the Spanish EMS response. Two of the reports note weaknesses in planning, communication, and coordination that resulted, for example, in an unbalanced distribution of patients to hospitals. One hospital was overloaded with patients while others had unused capacity. Confusion was exacerbated by the nature and locations of the attacks (multiple trains/stations/sites), poor connectivity between EMS organizations and receiving hospitals as well as between the two major EMS organizations in Madrid (SAMUR and SUMA) that did not routinely function together.[7]

FIGURE 10.3 Photos of field triage, including the temporary field medical assistance post. (Courtesy of Dr. Ervigio Corral Torres; Presented at a Tale of Cities meeting, New York, March 15, 2011. Available at http://www.taleofourcities-newyork.com/presentations/Day1/Torres%20NY% 20March%202011.pdf.) (For a color version of this figure, see the color plate section.)

Other important aspects of the bombings included their psychological effects on the public, the role of bystanders in providing immediate assistance, the function of forensic specialists in identifying the dead, and addressing the many social needs, particularly among children and the elderly.

Table 10.1 summarizes issues related to public health, EMS, and hospital groups in preparing for and responding to terrorism related mass casualty events.

OTHER BOMBINGS AND GLOBAL LESSONS

London

In addition to the case study in Chapter 1 of this volume, there is much literature on the July 7, 2005, London bombings. It covers the immediate emergency response,[8] EMS,[9] the Royal London Hospital experience,[10] the functioning of the Royal London Hospital trauma center and the intensive care medical service,[11] and resilience and psychological reactions of the general population.[12] At a Tale of Cities meeting, Mr. Simon Lewis, then Chief Superintendent of Metropolitan Police Operations at New Scotland Yard, focused on additional critical issues relevant to public health and its interactions with other first

TABLE 10.1 Summary Planning Checklists for Terrorist Bombing Events

A. Expected Health System Challenges
 1. Leadership
 2. Prehospital care
 3. Patient transport and distribution
 4. Hospital care
 5. Community and media relations

B. Principles for Health System Preparedness in Emergencies
 1. Provide meta-leadership
 2. Decide who is in charge
 3. Be proactive and expect the unexpected
 4. Learn from others
 5. Exercise MCE response plans
 6. Involve the public: they are an asset not an obstacle
 7. Work effectively with the media: inform the public in a timely manner
 8. Develop connected emergency plans
 9. Communicate during an MCE
 10. Be prepared for legal and ethical issues
 11. Alter standards of care
 12. Develop resilient medical surge

C. Basic Principles for Prehospital Care during an MCE
 1. Maximize availability of emergency medical services personnel and resources
 2. Assess the situation and care required
 3. Protect on-scene personnel
 4. Stage and triage patients
 5. Provide appropriate transportation and primary and secondary distribution of patients
 6. Manage fatalities

D. Common Challenges for Hospitals
 1. Predicting patient inflow
 2. Delays in declaring an MCE
 3. Time constraints
 4. Limited healthcare workforce
 5. Poor triage

E. Components of Patient Surge Management
 1. Planning
 2. Surge capacity and capability map
 3. Exercises and drills
 4. Redundant systems
 5. Triage and level of care
 6. Hospital Incident Command System
 7. MCE sites
 8. Security
 9. Recovery: ending the emergency status

Source: Adapted from Interim Planning Guidance for Preparedness and Response to a Mass Casualty Event Resulting from Terrorist Use of Explosives, Centers for Disease Control and Prevention, 2010. MCE, mass casualty event.

responders, such as the London Resilience Partnership and response planning (www.leslp.gov.uk), the response command and control structure, the forensic investigations at the bomb sites, mortuary and forensic identification issues, and the humanitarian family assistance plan for relatives of victims.[13]

Oklahoma City

There is also considerable literature on the terrorist bombing of the Alfred P. Murrah Federal Building on April 19, 1995. It includes descriptions of the epidemiology and nature of the injuries, location of victims at the time of injury, the risk factors for injury such as flying glass and other debris and collapsed ceilings,[14] psychiatric disorders among survivors of the bombing (45% of survivors studied had a post-disaster psychiatric disorder, and 34% had post-traumatic stress disorder [PTSD]),[15] and a review of physical injuries to rescue workers following the bombing.[16]

Israel

Israel has had considerable experience with terrorist bombings. In the 2000–2006 period more than 20,000 terrorist attacks were attempted against Israelis that resulted in 1084 deaths and 7633 injured, mostly civilians. Israeli experts have published numerous articles on the clinical and epidemiologic aspects of these events.[17] As a result of this extensive experience, Israel's public health leaders have developed many insights about the thinking of a terrorist, the nature and purpose of an attack, effective leadership in responding to an attack, timely communication and dissemination of information, effective and rapid triage, optimal distribution of casualties by the EMS to appropriate medical facilities, and appreciation for rapid clean-up and restoration at an attack scene.

Dr. Isaac Ashkenazi, former Surgeon General of the Israel Home Front Command, has freely discussed these issues at Tale of Cities meetings and other venues. Two recent books describe in detail the medical and clinical experience gained in Israel and the mechanisms for coping with disasters in Israel.[18]

Pakistan, India, Iraq, Afghanistan

These countries also have experienced many terrorist bombings. In 2010, in Pakistan for example, 473 separate bombings resulted in 1547 dead and 3581 injured. Some 25 Pakistani cities and towns experienced 49 bomb blasts each killing at least five individuals.[19] Although the US military has reported on its extensive experience related to EMS, triage, clinical diagnosis, and treatment, relevant English language publications in most of these countries are limited.[20]

In sum, whether in Madrid, London, or elsewhere, public health has played a critical role in response efforts. In the United States, the Centers for Disease Control and Prevention (CDC), which is the country's national public health agency, has engaged in numerous ways: by providing guidance documents for proper response, establishing partnerships with key response groups, participating in government planning for preparedness and response, and developing acute care research agendas. The Tale of Cities meetings have become an especially useful international venue for information sharing on issues relevant to public health preparedness and response.

OVERVIEW OF TERRORISM/DISASTER SCENARIOS AND PUBLIC HEALTH RESPONSE

The most common terrorist weapon is an explosive device, but other forms of weaponry and other causes of disaster are no less significant. Moreover, they often require more complex and varied responses. The range of public health activities is reflected in the responses associated with several well-known disparate events:

Terrorism

World Trade Center, 2001   A primary role of public health is to assess the mortality and morbidity related to terrorism/disaster events. Measuring the magnitude of the health impact involves surveillance and epidemiology, and the establishment of long-term studies and/or registries.[21] In the 9/11 attack, the ratio of fatalities to injuries was much higher than in the explosive bombings discussed above, so there was a far greater relative effort focused on mortuary services and identification of the dead utilizing DNA and other identification techniques.[22]

Syndromic surveillance in this instance was unrevealing, and despite the fears of some, evidence of exposure to biological agents was absent. However, there was concern about exposure to asbestos, heavy metals, and other chemicals that were in the dust and debris. Accordingly, New York City health authorities with the assistance of the Agency for Toxic Substances and Disease Registry (ATSDR) and the CDC established a registry of 71,000 participants to track potential long-term effects. Studies to date have identified a substantial dose-related, increased risk of new onset asthma and of PTSD in this population. A number of epidemiologic investigations have also looked at potential health effects in specific populations including pregnant women, infants and children, those with preexisting illnesses, and those who resided nearby.[23]

Because of the intense fires created by the attack and the numerous firefighters engaged, the many first responders exposed to the dust clouds of the crumpling buildings, and the months-long removal of debris, occupational health has figured prominently in this event. Long-term study of the firefighters has been undertaken and ongoing health services have been provided at several leading area hospitals. They have identified numerous respiratory and mental health illnesses among the first responders and those who worked extensively on the debris pile.[24]

Environmental assessment has been another essential part of the public health response. The effort has been collaborative among the US Environmental Protection Agency (EPA) and other environmental health agencies. The uncontrolled and sudden destruction of two enormous skyscrapers has led to extensive investigations of the dispersion of the dust plume, efforts to identify and quantify contaminants in the dust, evaluations of potential hazards to nearby residents, and information to guide remediation activities throughout the area. In addition to the demanding work of the laboratory investigations, major efforts have been applied to addressing contentious issues raised by these investigations, through community education and scientific oversight.[25]

Natural Disasters

Hurricane Katrina, 2005   When facing a major hurricane, health officials are likely to work with other first responders on preparedness functions such as planning for population evacuations, conducting surveillance for mortality (often done through medical examiner offices) and for illness, and communicating about common hazards post-hurricane, such as the peril of carbon monoxide poisoning from inappropriate use of generators when power lines are disrupted.[26]

Katrina posed huge challenges to public health officials and required more extensive responses than did other hurricanes in recent years. Key issues included the widespread destruction it caused across multiple states, including the disruption of clinical services; the loss of electric, gas, transportation, and other services; devastating flooding that led ultimately to the evacuation of New Orleans; the need to provide shelter, subsistence, and medications for an extended period to one million evacuees; and the challenges in repopulating New Orleans, reestablishing its health department, and returning to basic public health functions.[27]

Public health officials also were heavily engaged, often in conjunction with environmental and other agencies, as part of the broader government-wide response, in multiple specific challenges.[28] They included assessment of enormous amounts of debris and decisions about safe disposal of solid and hazardous wastes, assessments regarding release of and exposure to hazardous chemicals from industrial facilities and waste sites,[29] education of first responders about safety and health issues post-flooding,[30] and investigation of infectious disease outbreaks.[31]

A guide to dealing with the myriad challenges post-Katrina would in itself constitute a textbook on public health. It would also be instructive about other disasters, such as major earthquakes, that are disruptive to large populations and destructive of society's ability to provide the spectrum of basic services.

Mount Saint Helens Volcanic Eruption, 1980   The Mount Saint Helens (MSH) eruption was a reminder that disasters and acts of terrorism may present unique challenges and require specialized roles for public health officials. This eruption led to key responses: analysis of volcanic ash over a wide area (such ash may contain high levels of toxic elements such as arsenic and fluoride, though this was not the case with MSH); surveillance for respiratory disease as a consequence of the ash fall (respiratory disease was found to be elevated, but primarily in those with preexisting asthma); surveillance for respiratory disease in certain occupational groups (such as loggers working on harvesting the blown down trees); extensive public communications about safety concerns (such as the danger of roof collapses from accumulations of ash); and assessment of damage to infrastructure (the deleterious impact of ash on machinery as in water treatment plants).[32] Over several months the CDC provided 23 MSH Health Bulletins on health related issues in the Pacific Northwest. Maintaining up-to-date information on hazards such as these on public health agency Web sites continues to be a CDC function.

Radiation, Chemical, and Biological Disasters/Terrorism

Apart from the broad public health roles already enumerated, lie specialized responsibilities of public health officials for radiation, chemical, and biological events.

Radiation: Three Mile Island, 1979; Chernobyl, 1986; Fukushima, 2011   Radiation science is highly complex and the terminology can sometimes seem baffling. However, the crucial message here is that public health has an important role in preparedness for and response to radiation events. Critical issues that will engage public health officials include

- *Environmental Assessment and Dose Estimation.* There are multiple aspects of this effort, including estimation of the amount and nature of radiation released from the source, measurement of radionuclides in the environment (including in food, water, plants, and animals), estimation of doses to exposed individuals, and measurement of radionuclides within exposed individuals. Medical treatment regimens and decisions such as evacuation will be based on these estimates and measurements. Key public health laboratories, such as those at the CDC, are developing expanded capacities for measuring multiple radionuclides in large numbers of exposed people.

- *Public Health Guidance.* At Three Mile Island, radiation estimates outside the plant did not reach levels that required a major evacuation, impoundment of food or milk, or medical treatment decisions. At Chernobyl and Fukushima, off-site levels were much higher leading to evacuation of populations, decisions about acceptability of food supplies, and extensive testing of individuals for potential exposure.

- *Occupational Health.* This became a major concern at all three plants, where workers and first responders experienced actual or potential prolonged radiation exposure.

- *Remediation of Contaminated Areas.* An underappreciated but important issue concerns the return of people to evacuated areas. This involves estimation of health risks from residual radionuclides and an understanding of environmental radiological health policies including the levels of risk that are deemed tolerable.

- *Long-term Health Studies and Registries.* At Three Mile Island, dose estimates indicated that the risk of contracting cancer or other health problems as a result of radiation exposure was minimal. Still, a registry of 35,000 people within a 5-mile radius of the plant was created, so that they could be easily reached if health questions arose in the future.[33] At Chernobyl, thyroid disease and cancer, particularly in children, later developed at elevated levels.[34] In Fukushima, where off-site doses were higher than at Three Mile Island, plans reportedly were underway months after the incident to investigate large numbers of children for possible thyroid effects.

These nuclear power plant incidents have all been helpful toward understanding the potential scope of a radiation event whether the source is a nuclear meltdown, detonation of a dirty bomb, or ultimately use of a nuclear weapon.

Chemical: Bhopal, 1984   The accidental release of methyl isocyanate (MIC) from a chemical plant in Bhopal, India, was perhaps the largest ever chemical disaster. Approximately 3000 people were killed, 100,000 suffered acute symptoms from exposure, and 250,000 were evacuated from their homes. The disaster was exacerbated by the proximity of the plant to a heavily populated area, lack of warning or evacuation procedures for the exposed population, and poor plant maintenance resulting in failed safety features. The event also prompted a huge surge of patients, which overwhelmed the local hospitals.

Many toxicology issues can arise during a chemical disaster. In this case, the possible long-term effects of the MIC release have been a major concern. Unfortunately, only limited studies were initiated and no registry was established, which reduced opportunities for later identification of health impact in the exposed population.[35] The Indian Council of Medical Research did recently establish an environmental research center in Bhopal to stimulate further investigations even though more than 25 years have passed since the event.

Biological: Anthrax, 2001   The anthrax attacks via the US mail represent yet another category—biological agents—with the potential to cause great harm. Agents that cause infectious disease, whether long-known such as anthrax and plague, or recently identified such as SARS, can have a huge public health impact. Indeed, the public health role in disease settings is more familiar than in several of the other settings mentioned above. Still, many public health functions are common to all including

- diagnosis, screening, and treatment
- emergency medical services and surge capacity in hospitals
- environmental assessment
- occupational health
- short- and long-term health studies
- surveillance
  What is distinctive about biological agents is their ability to cause infection (and in some cases contagious disease) and the need to handle them in specialized laboratories. Public health hallmarks of the anthrax episode include
- emphasis on the clinical characterization of cases and development of diagnostic and treatment guidance[36]
- demonstrated need for new approaches to planning, prevention, preparedness, and surveillance[37]
- the huge impact on microbiological laboratories and the development of the Laboratory Response Network[38]
- development of new laboratory diagnostic techniques[39]

The anthrax attacks also led to expanded bioterrorism research efforts within public health agencies, and to governmental reorganization to better prepare for and respond to bioterrorism events.[40]

RELEVANT PUBLIC HEALTH PROFESSIONALS AND PREPAREDNESS

As has been abundantly shown, public health responses to disasters are remarkably diverse and they engage many different professionals. Certain functions are essential to all disasters and terrorist events: organizational leadership (meta-leaders in this context),[41] logistics and support staff, patient tracking systems, communications staff to assure rapid dissemination of information to the public, technical staff experienced with an all-hazards response to emergencies. In addition, the skills of specialists may variously be needed—epidemiologists, microbiologists, toxicologists, health physicists. Preparedness in public health agencies therefore includes practicing together to assure that at the time of a crisis the responders will function well together.

Table 10.2 provides an overview of key responder professions, though some of the scenarios outlined above warranted other clinical specialists as well. Thus, when large-scale evacuations to shelters are necessary, as with Hurricane Katrina, many other specialists could be called to service including pediatricians, internists, nurses, pharmacists, dieticians, dentists, and veterinarians. Any large public health agency, such as the CDC, a state health department, or even a large city health department, must be prepared to engage dozens of different professions for an effective response. This requires careful planning to assure that all the potential participants are rehearsed and available when needed.

Drills and Exercises for Disaster Preparedness

In Israel, preparedness is widely perceived as a matter of life and death and is therefore an urgent priority. Drills and exercises there often go beyond the written plan, the long

TABLE 10.2   Public Health Professionals Engaged in Response to Disasters and Bioterrorism

Public health leaders; organization directors	May have diverse backgrounds, but leadership training and experience essential
Emergency staff necessary for all-hazards response	Assure operational and administrative response
	Maintain emergency operation centers and arrange deployment of responders
	Handle all administrative and financial needs
	Provide staff training and support
	Logistical staff for distribution of stockpiled drugs, equipment, antidotes, and diagnostic materials
Medical/clinical responders	Emergency medicine physicians, EMS staff, various medical specialties (such as surgeons, orthopedists), medics, nurses, physician assistants, etc.
	Search and rescue teams
	Field hospital personnel
Epidemiologic/statistical/ informatics investigators	MD, PhD, MPH, and other trained public health professionals to track cases, health impact, risk factors, and preventive actions and interventions; and conduct epidemiologic studies, surveillance, registries, and monitoring and evaluation
Environmental assessors	Engineers, environmental scientists, health physicists, hydrologists, geologists, plume modelers, toxicologists, risk assessors, laboratory scientists
Occupational/worker health responders	Occupational medicine physicians, industrial hygienists, safety experts, occupational clinic staff, training and communications staff
Death investigators	Medical examiners/coroners, mortuary staff, forensic specialists, DNA laboratory staff, dentists
Infectious disease/biologic responders	Infectious disease physicians, vaccine distributors, food and water safety experts, microbiologists, laboratory technicians, environmental microbiologists, hospital infection control staff
Mental health/population resilience providers	Psychologists, psychiatrists, social workers, nurses, behavioral scientists, clinic staff, health educators

document, and the discussion group. Rather the emphasis is on in-depth and practical experience during a simulated event.[42] Exercises should include all relevant leaders, personnel, and agencies. They should be as realistic as possible, sometimes unannounced, and sometimes include "smart" casualties (actors playing the part of the wounded).[43]

In the United States, the National Exercise Program (NEP) provides an organized approach to set priorities for exercises. The NEP, which is part of the Federal Emergency Management Agency (FEMA), reflects those priorities in a multi-year schedule of exercises that serves the strategic and policy goals of the US Government. See http://www.fema.gov/prepared/exercise.shtm

CONCLUSION

The role of public health in disaster/terrorist events is to prevent death, disease, injury, and disability, and to promote health, wellbeing, and resilience in the affected population. A multitude of functions and activities go into an effective response.

Summary points:

- The three main goals in the immediate phase of emergency response are saving lives, reducing morbidity and suffering, and increasing community resilience.
- Public health is a leading partner in achieving these three critical goals.
- Public health must be an integral part of the leadership in preparing for and responding to all disasters. This means effective linkages through the Emergency Operations Centers and field headquarters established for managing such events. In the United States this is accomplished within the structure of the Incident Command System under FEMA.
- To ensure a flexible, synergistic, and coordinated response, public health must work effectively with other first responder agencies; planning and preparedness are essential to assuring that this will work smoothly during crisis.
- Public health leaders must assure the appropriate response and optimal functioning of the public health system (all relevant agencies and departments) during the crisis.
- Emergency response is not just a function of government. Educating the public so that bystanders, the true "first responders," can help in the immediate phase is critical. Similarly, engaging the broader society through communication, trust, and transparency is necessary to create the broadest possible support throughout the emergency response.
- Every response must be followed by rigorous evaluation that will shape the next round of planning and response. This process should begin immediately and the feedback loop must lead to rapid incorporation of lessons learned.
- The resilience of the society—enduring the disaster to overcome, persevere, and carry-on—is of paramount importance. The quality of the response can help promote resilience and enable the society to cope, recover, and rebuild.

NOTES

1. Turegano-Fuentes F, Caba-Doussoux P, Jover-Navalon JM, et al. Injury patterns from major urban terrorist bombings in trains: The Madrid experience. World J Surg 2008;32:1168–1175. Peral Gutierrez de Ceballos J, Turegano-Fuentes F, Perez Diaz D, et al. Casualties treated at the closest hospital in the Madrid, March 11, terrorist bombings. Crit Care Med 2005;33:107–112; 33(Suppl):S107–S112. Peral Gutierrez de Ceballos J, Turegano-Fuentes F, Perez-Diaz D, et al. 11 March 2004: The terrorist bomb explosions in Madrid, Spain—An analysis of the logistics, injuries sustained and clinical management of casualties treated at the closest hospital. Crit Care 2004; DOI: 10.1186/cc2995.

2. Miquel Gomez A, Jimenez Dominguez C, Ibarguren Pedrueza C, et al. Management and analysis of out-of-hospital health-related responses to simultaneous railway explosions in Madrid, Spain. Eur J Emerg Med 2007;14:247–255. Lopez Carresi A. The 2004 Madrid train bombings: An analysis of pre-hospital management. Disasters 2008; DOI: 10.1111/j.0361-3666.2007.01026.x; 41–65. Bolling R, Ehrlin Y, Forsberg R, et al. KAMEDO Report 90: Terrorist attacks in Madrid, Spain, 2004. Prehosp Disast Med 2007;22:252–257.

3. Turegano-Fuentes et al.; Peral Gutierrez de Ceballos et al. Casualties treated at the closest hospital; Peral Gutierrez de Ceballos et al. The terrorist bomb explosions in Madrid.

4. DePalma RG, Burris DG, Champion HR, Hodgson MJ. Blast injuries. N Engl J Med 2005;352:1335–1342.

5. Turegano-Fuentes et al.

6. Hunt R, Ashkenazi I, Falk H. A tale of cities. Dis Med Public Health Preparedness 2011;5(Suppl 2):S185–S188.

7. Miquel Gomez et al.; Lopez Carresi A et al.; Bolling R et al.

8. Holden PJP. Improvising in an emergency. N Engl J Med 2005;353:541–543.

9. Ryan J, Montgomery H. Terrorism and the medical response. N Engl J Med 2005;353:543–545. Redhead J, Ward P, Batrick N. Prehospital and hospital care. N Engl J Med 2005;353:546–547. Lockey DJ, MacKenzie R, Redhead J, et al. London bombings July 2005: The immediate pre-hospital medical response. Resuscitation 2005;66:ix–xii.

10. Aylwin CJ, Konig TC, Brennan NR, et al. Reduction in critical mortality in urban mass casualty incidents: Analysis of triage, surge, and resource use after the London bombings on July 7, 2005. Lancet 2006;368:2219–2225.

11. Mohammed AB, Mann HA, Nawabi DH, et al. Impact of London's terrorist attacks on a major trauma center in London. Prehosp Disast Med 2006;21:340–344. Shirley PJ. Critical care delivery: The experience of a civilian terrorist attack. J R Army Med Corps 2006;152:17–21.

12. Wessely S. Victimhood and resilience. N Engl J Med 2005;353:548–550. Rubin GJ, Brewin CR, Greenberg N, et al. Enduring consequences of terrorism: 7-month follow-up survey of reactions to the bombings in London on 7 July 2005. Br J Psychiatry 2007;190:350–356. Page L, Rubin J, Amlot R, et al. Are Londoners prepared for an emergency? A longitudinal study following the London bombings. Biosecur Bioterror 2008;6:309–319. Catchpole MA, Morgan O. Physical health of members of the public who experienced terrorist bombings in London on 07 July 2005. Prehosp Disast Med 2010;25:139–144.

13. Hunt, Ashkenazi, Falk.

14. Mallonee S, Shariat S, Stennies G, et al. Physical injuries and fatalities resulting from the Oklahoma City bombing. JAMA 1996;276:382–387.

15. North CS, Nixon SJ, Shariat S, et al. Psychiatric disorders among survivors of the Oklahoma City bombing. JAMA 1999;282:755–762.

16. Dellinger AM, Waxweiler RJ, Mallonee S. Injuries to rescue workers following the Oklahoma City bombing. Am J Ind Med 1997;31:727–732.

17. Peleg K, Kellerman AL. Enhancing hospital surge capacity for mass casualty events. JAMA 2009;302:565–567. Adini B, Peleg K, Cohen R, Laor D. A national system for disseminating information on victims during mass casualty events. Disasters 2010;34:542–551. Peleg K, Jaffe DH; the Israel Trauma Group. Are injuries from terror and war similar? A comparison study of civilians and soldiers. Ann Surg 2010;252:363–369. Kosashvili Y, Loebenberg MI, Lin G, et al. Medical consequences of suicide bombing mass casualty incidents: The impact of explosion setting on injury patterns. Injury 2009;40:698–702. Shapira SC, Adatto-Levi R, Avitzour M, et al. Mortality in terrorist attacks: A unique modal of temporal death distribution. World J Surg 2006;30:2071–2077. Aschkenasy-Steuer G, Shamir M, Rivkind A, et al. Clinical review: The Israeli experience: Conventional terrorism and critical care. Crit Care 2005;9: 490–499.

18. Shapira SC, Hammond JS, Cole LA, editors. *Essentials of Terror Medicine*. New York: Springer; 2009. Cole LA. *Terror: How Israel Has Coped and What America Can Learn*. Bloomington, IN: Indiana University Press; 2007.

19. Personal communication to Dr. Isaac Ashkenazi from Dr. Rashid Jooma, former Director General of Health Services in Pakistan, March 15, 2011.

20. MacDonald CL, Johnson AM, Cooper D, et al. Detection of blast-related traumatic brain injury in US military personnel. N Engl J Med 2011;364:2091–2100. Deshpande AA, Mehta S, Kshirsagar NA. Hospital management of Mumbai train blast victims. Lancet 2007;369:639–640.

21. Centers for Disease Control and Prevention. New York City Department of Health response to terrorist attack, September 11, 2001. MMWR 2001;50:821–822. Gibbs L, Farley T; World Trade

Center Medical Working Group of New York City. 2011 Annual Report on 9/11 Health. New York City, 2011. Available at www.nyc.gov/9-11HealthInfo

22. Biesecker LG, Bailey-Wilson JE, Ballantyne J, et al. DNA identifications after the 9/11 World Trade Center attack. Science 2005;310:1122–1123.

23. Farfel M, DiGrande L, Brackbill R, et al. An overview of 9/11 experiences and respiratory and mental health conditions among World Trade Center Health Registry enrollees. J Urban Health 2008;85:880–909. Wheeler K, McKelvey W, Thorpe L, et al. Asthma diagnosed after 11 September 2001 among rescue and recovery workers: Findings from the World Trade Center Health Registry. Environ Health Perspect 2007;115:1584–1590. Galea S, Ahern J, Resnick H, et al. Psychological sequelae of the September 11 terrorist attacks in New York City. N Engl J Med 2002;346:982–987.

24. Prezant DJ, Weiden M, Banauch GI, et al. Cough and bronchial responsiveness in firefighters at the World Trade Center site. N Engl J Med 2002;347:806–815. Banauch GI, Hall C, Weiden M, et al. Pulmonary function after exposure to the World Trade Center collapse in the New York City Fire Department. Am J Resp Crit Care Med 2006;174:312–319.

25. Lioy PJ, Weisel CP, Millette JR, et al. Characterization of the dust/smoke aerosol that settled east of the World Trade Center (WTC) in Lower Manhattan after the collapse of the WTC 11 September 2001. Environ Health Perspect 2002;110:703–714. Centers for Disease Control and Prevention. Potential exposures to airborne and settled surface dust in residential areas of Lower Manhattan following the collapse of the World Trade Center—New York City, November 4–December 11, 2001. MMWR 2003;52:131–136.

26. Falk H. Environmental health in MMWR—1961–2010. MMWR 2011;60(Suppl):S86–S96. Falk H, Briss P. Environmental- and injury-related epidemic-assistance investigations, 1946–2005. Am J Epidemiol 2011;174(Suppl):S65–S79.

27. Falk H, Baldwin G. Environmental health and Hurricane Katrina. Environ Health Perspect 2006;114:A12–A13.

28. Centers for Disease Control and Prevention. Public health response to Hurricanes Katrina and Rita—Louisiana, 2005. MMWR 2006; 55: 29–30. Centers for Disease Control and Prevention. Public health response to Hurricanes Katrina and Rita—United States, 2005. MMWR 2006;55:229–231.

29. Joint Task Force: Centers for Disease Control and Prevention and US Environmental Protection Agency. Environmental health needs and habitability assessment. Atlanta, GA, 2005. Available at http://www.epa.gov/katrina/reports/envneeds_hab_assessment.html, accessed June 20, 2012.

30. Centers for Disease Control and Prevention. Health hazard evaluation of police officers and firefighters after Hurricane Katrina—New Orleans, Louisiana, October 17–28 and November 30–December 5, 2005. MMWR 2006;55:456–458.

31. Yee EL, Palacio H, Atmar RL, et al. Widespread outbreak of norovirus gastroenteritis among evacuees of Hurricane Katrina residing in a large "megashelter" in Houston, Texas: Lessons learned for prevention. Clin Infect Dis 2007;44:1032–1039.

32. Baxter PJ, Ing R, Falk H, et al. Mount St Helens eruptions, May 18 to June 12, 1980: An overview of the acute health impact. JAMA 1981;246:2585–2589. Baxter PJ, Ing R, Falk H, et al. Mount St Helens eruptions: The acute respiratory effects of volcanic ash in a North American community. Arch Environ Health 1983;38:138–143. Buist AS, Bernstein RS, editors. Health effects of volcanoes: An approach to evaluating the health effects of an environmental hazard. Am J Public Health 1986;76(Suppl):S1–S90.

33. Talbott EO, Youk AO, McHugh-Pemu KP, Zborowski JV. Long-term follow-up of the residents of the Three Mile Island accident area: 1979–1998. Environ Health Perspect 2003;111:341–348.

34. Bard D, Verger P, Hubert P. Chernobyl, 10 years after: Health consequences. Epidemiol Rev 1997;19:187–204. Baverstock K, Williams D. The Chernobyl accident 20 years on: An

assessment of the health consequences and the international response. Environ Health Perspect 2006;114:1312–1317.

35. Dhara VR, Dhara R. The Union Carbide disaster in Bhopal: A review of health effects. Arch Environ Health 2002;57:391–404. Dhara VR. What ails the Bhopal disaster investigations? (And is there a cure?). Int J Occup Environ Health 2002;8:371–379. Koplan JP, Falk H, Green G. Public health lessons from the Bhopal chemical disaster. JAMA 1990;264:2795–2796.

36. Centers for Disease Control and Prevention. Investigation of bioterrorism-related anthrax and interim guidelines for clinical evaluation of persons with possible anthrax. MMWR 2001;50:941–948. Jernigan JA, Stephens DS, Ashford DA, et al. Bioterrorism-related inhalational anthrax: The first 10 cases reported in the United States. Emerg Infect Dis 2001;7:933–944.

37. Rotz LD, Hughes JM. Advances in detecting and responding to threats from bioterrorism and emerging infectious disease. Nat Med 2004;10(Suppl):S130–S136.

38. Trust for America's Health. Public health laboratories: Unprepared and overwhelmed. June 2003. Available at www.healthyamericans.org, accessed June 20, 2012.

39. Boyer A, Moura H, Woolfitt AR, et al. From the mouse to the mass spectrometer: Detection and differentiation of the endoproteinase activities of botulinum neurotoxins A-G by mass spectrometry. Anal Chem 2005;77:3916–3924.

40. Trust for America's Health. Remembering 9/11 and anthrax: Public health's vital role in national defense. September 2011. Available at www.healthyamericans.org, accessed June 20, 2012.

41. Marcus LJ, Dorn BC, Henderson JM. Meta-leadership and national emergency preparedness: A model to build government connectivity. Biosecur Bioterror 2006;4:128–134.

42. Pinkert M, Lehavi O, Benin Goren O, et al. Primary triage, evacuation priorities, and rapid primary distribution between adjacent hospitals—Lessons learned from a suicide bomber attack in downtown Tel Aviv. Prehosp Disaster Med 2008;23:337–341. Almogy G, Belzberg H, Mintz Y, et al. Suicide bombing attacks, update and modification to the protocol. Ann Surg 2004;239:295–303. Lynn M, Gurr D, Memon A, Kaliff J. Management of conventional mass casualty incidents: Ten commandments for hospital planning. J Burn Care Res 2006;27:649–658. Einav S, Feigenberg Z, Weissman C, et al. Evacuation priorities in mass casualty terror-related events—Implications for contingency planning. Ann Surg 2004;239:304–310. Hirshberg A. Multiple casualty incidents—Lessons from the front line. Ann Surg 2004;239:322–324.

43. Ashkenazi I, Mahany M. Hospital preparedness for emergencies. In: Suresh D, Brown AFT, Nelson B, Banarjee A, Anantharaman V, editors. *Textbook of Emergency Medicine*. New Delhi, India: Wolters Kluwer; 2011. National Center for Injury Prevention and Control. Interim planning guidance for preparedness and response to a mass casualty event resulting from terrorist use of explosives. Atlanta, GA: Centers for Disease Control and Prevention; 2010.

11

THE ROLE OF THE HOSPITAL RECEIVER

Mark A. Merlin

Whether for a drill or a real incident, emergency medical services (EMS) physicians typically are present at a hospital site or involved by telecommunication. One potential role for every EMS doctor is that of hospital receiver of patients from a terror or disaster event. While other personnel might also serve in that role, EMS physicians' qualifications are enhanced by their multi-varied responsibilities.

EARTHQUAKE, HURRICANE, FLOOD

As one who has been involved in EMS and disaster preparedness for more than 25 years, a vacation is often different for me than for people in other fields. My associates know that I will still be available "24/7." On August 20, 2011, I departed Newark airport for a 10-day California vacation. Although the West Coast is prone to earthquakes and other natural disasters, the possibility of encountering a mass casualty incident (MCI) did not cross my mind. Three days later an earthquake erupted, though ironically not in California, but on the East Coast. The epicenter was in Virginia. Slight tremors were felt hundreds of miles away including in New Jersey, where I serve as medical director of the New Jersey EMS Task Force.

During the first few hours, I received 36 emails. Almost all of them would have scared me if I had not understood that the epicenter was far my home state. Still, concerns were widespread and I began to consider whether I should return home early. I felt frustrated that I was so far from the event as I tried to remain continuously informed.

The information coming to me was unconfirmed, but it included possible dire mass casualty scenarios. Finally an email arrived from a trusted source confirming that the earthquake was minimally disruptive. Within 2 days, worries about the earthquake had

Local Planning for Terror and Disaster: From Bioterrorism to Earthquakes, First Edition.
Edited by Leonard A. Cole and Nancy D. Connell.
© 2012 John Wiley & Sons, Inc. Published 2012 by John Wiley & Sons, Inc.

faded, but new reports were appearing about another threat to the East Coast, Hurricane Irene. Conference calls about the impending hurricane started on Thursday, August 25, while I was still on "vacation." The first call included county coordinators and members of state agencies. Experts were warning that the storm could become the most devastating in New Jersey's history.[1]

On Friday the conferences calls continued and state plans were being reviewed. I decided to return immediately and was able to fly out that afternoon. For 6 hours in the air, I kept wondering what was happening on the ground in New Jersey. Upon landing in Newark, my phone indicated that I had received 72 new emails during the flight. Planning was continuing, though in the face of uncertainty about the impact of Irene. By Friday night I was speaking with available EMS physicians in the state who could provide medical direction perhaps at a damage site or as a hospital receiver.

The next morning brought heavy rain and evacuations began in southern parts of the state. Groups of people, including nursing home patients, were moved to facilities in the central part of the state. By evening the rain volume had increased. Wind speed reached 50 miles per hour, though remained lower than worst-case predictions of 120 miles per hour. But the rain continued to fall and rivers rose to record levels. More than 30,000 people were evacuated as flooding became widespread. Ten New Jerseyans died including Michael Kenwood, a medical rescue worker who was swept away by flood waters while trying to reach a submerged car.[2]

In the mist of the crisis I received a call from a shelter that desperately needed help with evacuees requiring medical support. Many had left home without their medications. Several people in the shelter were oxygen dependent and oxygen supplies were getting low as were food stocks. After several calls I was able to confirm that medications and food were being delivered from local hospitals.

Nurses and physicians had been in the shelters for 30 hours and were suffering from exhaustion. Some were unable to continue working. At one point, because of the diminished support staff, 35 patients had to be transferred to local hospitals. But after the acute phase of the hurricane had passed, many in the shelters were able to return to their homes or other facilities not affected by the flood.

In hindsight, the event prompted me to consider two key challenges. The first involves the experience with the shelters. The second relates to what would have been required if massive hospitalizations had been necessary, and particularly the functions of the hospital receiver.

Preparedness training for MCIs ordinarily includes little about the management of shelters. It is clear from issues that surfaced during Hurricane Irene that more attention to this area is necessary. Questions that need to be addressed include

- Who is responsible for providing medical and support care for those who have been evacuated to shelters?
- Who will provide the medications that might be required for prolonged periods?
- Who will provide the evacuees with transportation back to their residences or nursing homes?

A strategy to address these and other relevant challenges needs to be developed, rehearsed, and in place before the next MCI. Failure to provide minimally necessary resources for the care of patients in shelters could result in many more going to hospital emergency departments (EDs). This could have unwelcome consequences. A large patient influx could overwhelm a hospital's ability to deliver a high standard of care.

During and immediately after Hurricane Irene, the number of admissions to hospitals in New Jersey remained manageable. While the lives of many people were disrupted, few of them suffered injuries requiring hospitalization. This was doubly fortunate because hospitals themselves were short of medical staff during that period. The shortage would only have exacerbated the challenge of dealing with a massive influx of patients. At the forefront of the challenge would have been the hospital receiver.

PREPAREDNESS AND THE FIRST RECEIVER

Preparation to receive patients begins with proper notification to acute care facilities. Well thought out emergency plans and exercises lead indisputably to better preparedness.[3] Although many patients after an event walk into a hospital on their own, proper notification from the scene to the receiving centers is essential. Numerous questions that pertain to the hospital receiver should be answered in the planning stages:

- Upon arrival at the hospital, where are the new patients going to go?
- Should disaster victims be mixed with regular ED patients?
- Should disaster patients be brought to the ED or to another area of the hospital?
- Should walk-in patients be separated from those coming by ambulance?
- Will the victims be triaged through the same area as the regular ED patients?
- Are all the front-line caregivers familiar with decontamination procedures?
- How many patients can be decontaminated and in what time period?
- Is the hospital prepared, if necessary, to evacuate patients?
- What is the notification system if the event happens in the middle of the night?
- How quickly can, say, 100 nurses or physicians be activated?
- What is the method to recall general staff? (One study demonstrated that only 64% of the hospitals surveyed had recall protocols and only 56% of service chiefs knew the protocol.[4])
- How quickly can appropriate though diverse departments be reached, such as respiratory therapy, pathology, morgue, pastoral care, orthopedics, trauma, and radiology?
- Do contracts exist for water supply? Additional fuel? Emergency repair of damaged roofs or other infrastructure?
- How will hospital personnel communicate with blast victims who have lost hearing because of ruptured eardrums?
- Does the hospital have memoranda of understanding (MOU) with regional specialty centers?
- Do residents and rotating medical students have to read and take a quiz on the code triage manual?
- If the phone and Internet systems fail, how will personnel be contacted?
- What are the surge capacities of the ED, operating room, hemodialysis, other individual units?

- What temperatures in the facility must be reached for full-scale evacuation to be required?
- Where is all of this information kept, and do nightshift personnel know how to access it quickly?
- How does the hospital receiver work with the physicians discharging patients? This applies specifically to a situation when new patients cannot enter the system unless others are leaving it. (A recent study showed that hospitals could raise their surge capacity possibly by 68%, by identifying patients who potentially could be discharged early.[5])
- At what point during an unfolding event might outside resources be required? In July 2005, for example, the Royal London Hospital received 194 casualties from the London bombings. The hospital's capacity was reached in 15 minutes, after 17 patients were found to need immediate surgery and 265 units of blood.[6]

A review of the literature found that in the 5-year period after September 11, 2001, the number of articles on disaster-related topics in sub-specialty journals increased by 320%.[7] Still, few well-designed trials were published regarding the roles, responsibilities, and training of the hospital receiver. Thus comprehensive guidelines for the first receiver continue to await more evidence-based study.

ACUTE CARE HOSPITAL READINESS

In March 2011, the US Department of Health and Human Services released a report based on National Hospital Ambulatory Medical Care Surveys (NHAMCS).[8] The results were obtained from on-site representatives of the US Census Bureau, which offered an overview of acute care hospital preparedness. More than 90% of hospitals had general response plans for natural disasters and for biological and chemical events. Eighty-nine percent of hospitals had developed cooperative plans with EMS, though only 64% had cooperative plans with local hazardous materials teams. Fewer than 50% of hospitals had plans for sheltering special needs populations such as blind, deaf, mobility-impaired, or mentally challenged patients. Plans for pediatric patients were of special concern with only 29% of hospitals prepared to acquire supplies for children and 34% having plans for reuniting children with parents.[9]

Another study reported that fewer than half of Canadian trauma centers felt they were ready for an MCI.[10] Most communication plans involved cell phones or landlines even though these networks might be overwhelmed with callers during an event. Furthermore, callback systems for staff members not on call were largely inadequate.[11]

THE HOSPITAL RECEIVER: DEFINITION AND ROLES

Depending on availability and training, a hospital first receiver could be any one of several professionals including a physician, nurse, administrator, resident, x-ray technician, phlebotomist, respiratory therapist, or transporter. Hospitals are usually located at some distance from the site of a terror or disaster event. Thus exposure of hospital receivers to event-related substances are limited to whatever might be transported to the hospital on a victim's skin, hair, clothing, or personal effects.[12]

The location of the first receivers is an important distinction from that of first responders such as firefighters, law enforcement, and EMS professionals, who appear at the scene of the incident. Since the receivers are not at the scene, their exposure to patients can be better controlled and limited with proper training. The role of the first receiver includes establishing a post-decontamination zone that begins at the entrance to the hospital and is used to prevent further contamination.

The responsibilities of the hospital receiver should be outlined in a hospital's incident command system (HICS). The system generally is drawn from standard incident command system plans, but then placed in a hospital setting.[13] The HICS assists hospitals in emergency management planning, response, and recovery for unplanned and planned events. HICS or a similar structure should be activated during an emergency event or exercise. The advantage of HICS is that it is an established well-organized format by experts in the field. It also offers a strategy for dealing with incidents by all hospital personnel and interacting with the community instead of hospital preparedness in isolation.

An HICS calls for the establishment of a hospital command center (HCC). The HCC should be activated by the incident commander or administrator at the time of HICS activation. Although hospitals can create individualized forms for events, HICS is a valuable tool and training process for the first receiver. HICS form 207 represents an organization chart for all hospital personnel who potentially are first receivers (Figure 11.1).

Additional guidelines are provided by the Occupational Safety and Health Administration (OSHA). The roles of emergency first responders are covered under OSHA's Standard on Hazardous Waste Operations and Emergency Response (HAZWOPER). OSHA also recognizes that first receivers have different training and personal protective equipment (PPE) needs than workers in the hazardous substance zone. In a 2005 publication, OSHA explicitly addressed the subject of hospital-based first receivers. Although the focus was on best practices for receivers of victims exposed to hazardous substances, some of the practices may apply to receiving victims in general.[14]

The OSHA report provides a list of assumptions regarding communications, resources, and victims that a hospital and its receivers should anticipate in preparing for an MCI[15]:

- Victims will arrive with little or no warning to the hospital.
- Information regarding the hazardous agent(s) will not be available immediately.
- A large number of victims will be self-referred (as much as 80% of the total number).
- Victims will not necessarily have been decontaminated before arriving at the hospital.
- A high percentage of people arriving at the hospital will have experienced little or no exposure, which should be considered in decontamination plans.
- Most victims will go to the hospital closest to the site where the emergency occurred.
- Victims will use other entrances in addition to the ED.

In this best practices document, OSHA provides information to assist hospitals to deal with employee protection and training. While offering guidance to hospitals and hospital receivers, the document also references the Joint Commission on Accreditation of Healthcare Organizations (JCAHO). JCAHO suggests an "all-hazards approach" that permits flexibility in responding to emergencies of all types. These concepts allow hospital personnel to assume various responsibilities and allow flexibility depending on the type of MCI.

ORGANIZATION CHART

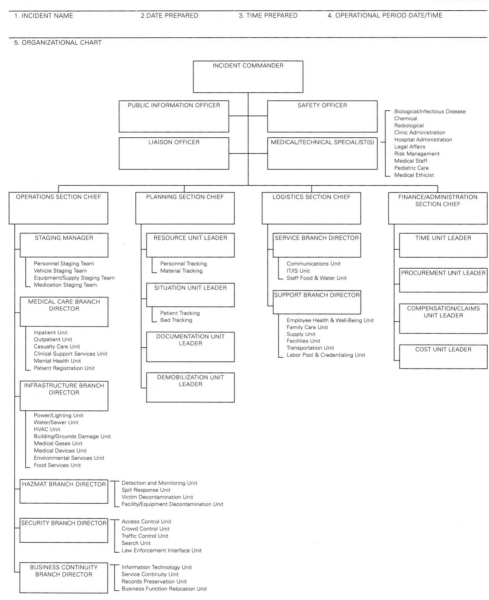

FIGURE 11.1   Organization chart for all members of the incident team (HICS Form 207).

TABLE 11.1   Minimum PPE for Hospital-based First Receivers of Victims from MCIs Involving the Release of Unknown Hazardous Substances

Scope and Limitations
This table applies when
   The hospital is not the release site
Prerequisite conditions of hospital eligibility are
   already met
The identity of the hazardous substance is unknown

Zone	Minimum PPE
Hospital Decontamination Zone	
All employees in this zone (includes, but not limited to, any of the following employees: decontamination team members, clinicians, setup crew, cleanup crew, security staff, and patient-tracking clerks)	PAPR that provides a protection factor of 1000; the respirator must be NIOSH-approved
	Combination: 99.97% high-efficiency particulate air (HEPA)/organic vapor/acid gas respirator cartridges (also NIOSH-approved)
	Double-layer protective gloves
	Chemical resistant suit
	Head covering and eye/face protection (if not part of the respirator)
	Chemical-protective boots
	Suit openings sealed with tape
Hospital Post-decontamination Zone	
All employees in this zone	Normal work clothes and PPE, as necessary, for infection control purposes (e.g., gloves, gown, appropriate respirator)

Note: This table is part of, and intended to be used with, the document entitled "OSHA best practices for hospital-based first receivers of victims from mass casualty incidents involving the release of hazardous substances."

Employee training and PPE selection processes are also defined in OSHA documents.[16] PPE selection should be based on a hazard assessment. Use of powered air-purifying respirators (PAPRs) is a practical form of respiratory protection for first receivers in the hospital decontamination zone. Since many MCIs do not require PAPRs, periodic retraining is necessary for personnel who might at some time be called upon to use these powered respirators. OSHA requires level B protection or self-contained breathing apparatus (SCBA) for unknown hazards. (Level B means that the highest level of respiratory protection is necessary but a lesser level of skin protection is needed.) However, it should be noted that SCBAs are cumbersome and can make caring for patients more difficult.

Table 11.1 shows OSHA's best practices for first receiver exposure and is essential for PPE training.

A higher level of protection also may be necessary for any hospital that anticipates providing specialized services (such as a Hazardous Materials Response Team at the incident site). OSHA requirements state, "If a hospital is responding to a known hazard, the hospital must ensure that the selected PPE adequately protects the employees from the identified hazard."

THE RECEIVER ROLE OF THE HOSPITAL

All hospitals must have an emergency management plan (EMP). While no institution can prepare for every emergency, hospitals require information that will help hospital receivers make informed decisions about the severity of the incident and the possibility of further exposure to staff. A hazard vulnerability analysis (HVA) assists hospitals in determining which PPE would be required for various scenarios. Specifically, the HVA ranks vulnerability to hazards and prioritizes efforts to reduce vulnerability. JCAHO mandates an HVA as part of emergency preparation.[17,18] An HVA should be reassessed on a routine basis or whenever severe weather conditions or other high-risk events dictate. In performing an adequate HVA it is essential that all members of the hospital preparedness team participate.

The HAZWOPER standard requires that hospital employees at the first responder operations level have a minimum of 8 hours of training or sufficient experience to demonstrate competency in the following areas[19]:

- An understanding of what hazardous substances are, and the risks associated with them in an incident
- An understanding of the potential outcomes associated with an emergency when hazardous substances are present
- The ability to recognize the presence of hazardous substances in an emergency through signs and symptoms of exposure
- The ability to identify the hazardous substances, if possible
- An understanding of their role in the hospital's emergency response plan, including site security and control, and decontamination procedures
- The ability to realize the need for additional resources and to make appropriate notifications to the communication center
- Knowledge of the basic hazard and risk assessment techniques
- Know how to select and use proper PPE
- An understanding of basic hazardous materials terms
- Know how to perform basic control, containment, and/or confinement operations within the capabilities of the resources and PPE available
- Know how to implement basic decontamination procedures
- An understanding of the relevant standard operating procedures and termination procedures

Regulation 29 CFR 1910.120 specifically deals with internal release of hazardous substances within a hospital or when a hospital is requested to send personnel to a scene of a large-scale event. The hospital must decide through preplanning which personnel will go to the scene and which will be retained to receive incoming patients. Although first receivers do not usually perform the duties of a first responder, their training should prepare them to assist a responder as needed.

Physical Responsibilities of the Hospital Receiver

When first receivers' activities require PPE the resulting physical stress exceeds that of normal daily routine. Accordingly, hospital medical staff should monitor employees' vital signs as one method of tracking employee response to these stressors. Hospitals should monitor weight, time working in PPE, and last meal of the receiver. This should be done on a regular basis during drills and real events. Hospitals must standardize documentation of employee performances during an event or drill. Typically, this task is given to the assigned safety officer in conjunction with the medical team. The length of work periods must be adjusted based on external temperature and the level of protection worn. Thermal stress can affect how long first receivers may perform their duties.

Decontamination procedures can have a large impact on first receiver exposure to hazardous substances. All the hospitals interviewed for the OSHA 2005 hospital receiver document agreed that the basic steps include[20]

1. Activate the EMP.
2. Learn as much as possible (as soon as possible) about the number of victims, the contaminant, and associated symptoms. Previous arrangements with first responder organizations can improve the timeliness and quantity of information received.
3. Activate the decontamination system and assemble the decontamination team and site security staff.
4. Perform any medical monitoring (e.g., vital signs), if specified by the EMP.
5. Put on PPE.
6. Triage victims to determine which individuals require decontamination and provide critical medical treatment to stabilize them before decontamination (e.g., atropine).
7. Assist victims (ambulatory and non-ambulatory) in removing contaminated clothing and securing personal property as soon as possible (within minutes of arrival).
8. Place clothing and other contaminated items in an approved hazardous waste container that is isolated outdoors so the items are not a continuing source of exposure.
9. Wash victims with soap, good surfactant properties, and water (preferably tepid water to improve victim compliance). This step should include copious rinsing.
10. Inspect victims to evaluate the effectiveness of decontamination and guide decontaminated victims to the medical treatment area (hospital post-decontamination zone). Return inadequately decontaminated victims to the shower area and repeat cleansing.
11. Decontaminate equipment and the decontamination system (if not disposable).
12. Staffs remove PPE and decontaminate themselves.

Prehospital to Hospital Communication

Communication is always a challenge during an MCI. It is especially important for hospital receivers, as gatekeepers, to maintain communication lines with other key personnel. Their means of communication should include not only radios, computers, and cell phones but also predefined messengers with paper in case the mechanical systems go down.

Leaders of departments must have access to these devices 24/7 and, when unavailable, they must assign these devices to someone who has an administrative position in the same department. Ultimately during an MCI, people need the ability to communicate rapidly and to make unilateral decisions.

Hospital communication starts with information in and out of a structured command system. It deals with a myriad of issues including those concerning security, patients, EMS, other hospitals, federal agencies, state agencies, family members, and the media. Initial information ideally is received before the first patient arrives. Proper integration of the information depends on long-standing working and drilling relationships among leaders of various hospital groups: physicians, nurses, administrators, and technicians. If these groups do not work well together on a daily basis, communications will more likely break down when the system is taxed.

Communicating with prehospital personnel begins with identification of surge capacity. The hospital receiver must be familiar with how many patients can be received and with any specialty capabilities of the institution. One caveat: The term surge capacity has various interpretations. A review of annual bed statistics in New Jersey hospitals demonstrated no limitations on reported surge capacity. Yet when daily bed statistics were considered, capacity was severely strained on Tuesdays, Wednesdays, Thursdays, and Fridays. Thus, on 288 days per year, presumptions about the hospitals' surge capacity did not match the reality.[21]

Like any other skill, successful preparedness depends on repeated practice. Tabletop exercises remain the most common methods of practice because they are effective and inexpensive. In fact, these exercises have exposed weaknesses at various times among key personnel including hospital administrators, physicians, and nurses.[22]

Studies also have highlighted other common failures including that of senior leadership to make rapid decisions,[23] and of staff to routinely update patients' charts, resulting in faulty treatment for worsening medical conditions.[24] These lapses suggest adopting a military-type strategy with clear delineation of leadership roles and trained individuals empowered to make unilateral decisions.

MEDICAL STAFF TRAINING

OSHA's best practices document recommends two levels of training: operations and awareness.[25] Operations education is specific for personnel working in a decontamination zone. Awareness level is appropriate for personnel who do not have specific duties in the decontamination zone but may encounter victims. Both categories should be addressed in a hospital's EMP.

Under normal circumstances, the concept of training a few individuals in various specialty areas has proved successful. We often call on specialists or subspecialists to take care of specific patient care issues. However, preparation for MCIs is different. Such events, which are often unexpected, may require an "all hands on deck" approach. Moreover, the nature of an MCI might not be clear at first, and only belatedly would it be recognized which resources and how many support people are needed. Thus, a degree of training is advisable for everyone who might serve in a support role. But since training can be expensive and time consuming, the question remains: How much training should be given to preparation for unlikely events?

Studies that address this question typically are based on small sample size without valid scoring systems. Still, a plethora of reports following drills and events identify many common lapses in addressing an MCI. They include

- deficiencies in chain of command
- poor nursing triage training
- delays in critical patient care

- poor medical training of police and fire assisting hospital staff
- significant over and under triage
- out-of-town ambulances' inability to find hospitals due to poor highway signs[26-34]

What is clear is that physicians and nurses without mass casualty training are limited in their ability to react appropriately in mass casualty events.[35] This is certainly true as well for other support staff whether full or part time, whether students in extended training programs or rotating through the hospital for brief periods. Relevant courses are now readily available. They include such titles as Basic Disaster Life Support, Advanced Disaster Life Support, and Incident Command Systems.

Training personnel for MCI scenarios can be enhanced by technology-based instruction. For example, computer-assisted modeling has been used to teach subjects ranging from treatment for burns and radiation exposure to information retention and crowd control.[36-39]

Familiarity with these topics could prove invaluable in the event of a terror or disaster incident. A hospital receiver who has trained in these areas would be better able to infer a patient's immediate needs, more expertly perform triage, and designate an appropriate receiving department.

Layered into a hospital receiver's assessments must be awareness of the hospital's surge capacity. An understanding of this capacity should extend to the ED, operating facilities, pediatrics, intensive care units, morgue, hemodialysis, and blood bank. In addition, hospitals should make local EMS aware of surge capacity for various departments. A retrospective study of hospital MCIs, defined as greater than 10 casualties sent to a single hospital, demonstrated 79% of patients did not need immediate medical treatment.[40] Thus, these patients could have been brought to an alternate site or treated and released at the scene.

Although most victims of MCIs are brought to a hospital, the concept of bringing more patients to alternative off-site locations should be explored. This approach might be preferable if no hospital is near the disaster scene or if nearby hospitals are overcrowded. Whichever the case, rehearsed interactions between hospital receivers and first responders could help to redistribute these patients.

Among the most important concerns of hospital receivers is the periodic shortage of blood and blood products. In the fall of 2010, for example, the Red Cross advised that hospitals in Los Angeles, Atlanta, and Philadelphia cancel elective surgery because of a blood shortage. A spokesperson for the organization said that 80,000 units should be on hand daily, but that the blood bank had only 36,000 units.[41] A year later, Hurricane Irene prompted a depletion of blood supplies in locations along the East Coast. Because of closed donor centers and canceled blood drives, New York supplies had dropped "well below the normal 3- to 5-day inventory level."[42]

An ample blood supply is indispensable to proper preparedness, especially for a hospital receiver's ability to function optimally. Thus, a blood bank representative should be at the table for preplanning events and exercises. These personnel are essential to logistical planning.

Finally, some special needs populations need singular attention during MCIs. Chief among them are dialysis patients. They require several hours of supervised life-sustaining treatment each week. If outpatient dialysis centers are involved in an MCI, these patients will likely present to a hospital. Thus, hospitals must be aware of their own limited dialysis capabilities, from the availability of machines to that of dialysis nurses. Advance preparedness is key to successful care.

A recent study evaluated patients after a dialysis outpatient center was significantly damaged during an earthquake in Italy. Eighty-eight patients were transferred in an orderly manner to other regional dialysis centers, and none missed any treatment sessions. Within 3 days, additional modified dialysis centers had been set up in tents. This successful outcome came about because of thoughtful planning and decisive leadership.[43]

CONCLUSION

Proper hospital preparedness for terror or disaster involves everyone affiliated with the facility. Drills and exercises should include personnel from all key disciplines. While every medical field is important to the system, the hospital receiver of patients holds a central position. Whether at the point of a patient's entry to the hospital or to a particular department, the designated receiver is an essential mover of the process.

During an MCI, the first receiver's role must be understood as distinct from that of the first responder's. Essential qualifications for the position include communication skills, availability beyond the standard workweek, and the ability to work effectively with other sectors of the hospital community.

NOTES

1. Sterling S. Hurricane Irene could be "most devastating storm ever to hit" N.J., climatologist says. Star Ledger, August 25, 2011.

2. Rojas C. Princeton rescue squad member dies from injuries sustained in Hurricane Irene rescue attempt. NJ.com [Trenton Times], August 29, 2011. Available at http://www.nj.com/mercer/index.ssf/2011/08/princeton_first-aider_dies.html, accessed September 20, 2011.

3. McAliter VC. Drills and exercises: The way to disaster preparedness. Can J Surg 2010;54(1):7–8.

4. Gebbie KM, Silber S, McCollum M, Lazar EJ. Activating physicians within a hospital emergency plan: A concept whose time has come? Am J Disaster Med 2007;2(2):74–80.

5. Kelen G, McCarthy ML, Kraus CK, et al. Creation of surge capacity by early discharge of hospitalized patients at low risk for untoward events. Disaster Med Public Health Prep 2009;3(2 Suppl):S10–S16.

6. Aylwin C, Konig TC, Brennan N. Reduction in critical mortality in urban mass casualty incidents; analysis of triage, surge and resource use after the London bombing on July 7, 2005. Lancet 2006;368(9354):2219–2225.

7. Kelen.

8. Niska R, Shimizu I. Hospital preparedness for emergency response: United States, 2008. Natl Health Stat Report 2011;37:1–15.

9. Ibid.

10. Gomez D, Haas B, Ahmed B. Disaster preparedness of Canadian trauma centres; the perspective of medical directors of trauma. Can J Surg 2010;54:9–16.

11. McAliter.

12. Horton DK, Orr M, Tsongas T, Leiker R, Kapil V. Secondary contamination of medical personnel, equipment, and facilities resulting from hazardous materials events, 2003–2006. Disaster Med Public Health Prep 2008;2(2):104–113.

13. Hospital Incident Command System, California Emergency Medical Services Authority, 2011. Available at http://www.emsa.ca.gov/HICS/, accessed September 20, 2011.

14. OSHA best practices for hospital-based first receivers of victims from mass casualty incidents involving the release of hazardous substances. Occupational Safety and Health Administration, 2005. Available at http://www.osha.gov/Publications/osha3249.pdf, accessed September 20, 2011.

15. Ibid.

16. Ibid.

17. Joint Commission on Accreditation of Healthcare Organizations. *Comprehensive Accreditation Manual for Hospitals: The Official Handbook.* Oakbrook Terrace, IL: Joint Commission Resources; 2003:EC–9.

18. Ibid; 2004/5.

19. Occupational Safety and Health Administration, 2005.

20. Ibid.

21. Delia D. Annual statistics give a misleading picture of hospital surge capacity. Ann Emerg Med 2006;4(48):384–388.

22. Gretenkort P, Harke H, Blazejak J. Interface between hospital and fire authorities—A concept for management of incidents in hospitals. Prehosp Disaster Med 2002;17(1):42–47.

23. Baughman KS, Calvert DR. Internal disaster drill critique. J Emerg Nurs 1990;16(4):240–241.

24. Lau PF, Lau CC. A disaster drill in Hong Kong. Accid Emerg Nurs 1997;5(1):34–38.

25. Occupational Safety and Health Administration, 2005.

26. Baughman.

27. Lau.

28. Eisner ME, Waxman K, Mason GR. Evaluation of possible patient survival in a mock airplane disaster. Am J Surg 1985;150(3):321–323.

29. Paris PM, Stewart RD, Pelton GH. Triage success in disasters: Dynamic victim-tracking cards. Am J Emerg Med 1985;3(4):323–326.

30. Saxena K, Cicero JJ, Hass WF. Chemical disaster drill in the State of Minnesota. Minn Med 1998;69(3):137–139.

31. Maxwell C, Cherneski M. Post-disaster drill analysis. Using victim-tracking cards. J Emerg Med Serv 1987;12(2):43–46.

32. Cook L. Hospital disaster drill game: A strategy for teaching disaster protocols to hospital staff 1990. J Emerg Nurs 1990;16(4):269–273.

33. Gofrit ON, Leibovici D, Shemer J. The efficacy of integrating "smart simulated casualties" in hospital disaster drills. Prehosp Disaster Med 1997;12(2):97–101.

34. Tur-Kaspa I, Lev EI, Hendler I. Preparing hospitals for toxicological mass casualties' events. Crit Care Med 1999;27(5):1004–1008.

35. Fishel ER. Exercise "Med-Ex" 73. Maryland State Med J 1994;23(9):46–48.

36. Levy K, Aghababian RV, Hirsch EF. An Internet-based exercise as a component of an overall training program addressing medical aspects of radiation emergency management. Prehosp Disaster Med 2000;15(2):18–25.

37. Gray D. Disaster plan education: How we made and tested a video. J Accid Emerg Med 1996;13(1):21–22.

38. Inglesby TV, Grossman R, O'Toole T. A plague on your city: Observations from TOPOFF. Clin Infect Dis 2001;32(3):436–445.

39. Burns KA. Experience in the use of gaming and simulation as an evaluation tool for nurses. J Contin Educ Nurs 1984;15(6):213–217.

40. Kosashvili Y. Israeli hospital preparedness for terrorism-related casualty incident: Can the surge capacity and injury severity distribution be better predicted? Injury 2009;40:727–731.

41. Zabarenko D. The nation has a major blood shortage. ABC News, September 19, 2010. Available at http://abcnews.go.com/Health/story?id=117954page=1, accessed September 18, 2011.

42. Irene causes blood shortage. Long Island Herald, September 1, 2011. Available at http://www.liherald.com/stories/Irene-caused-blood-shortage,35311, accessed September 18, 2011.

43. Bonomini M. Dialysis practice and patient outcome in the aftermath of the earthquake at L'Aguila, Italy, April 2009. Nephrol Dial Transplant 2011;26(8):2595–2603.

12

MANAGING TRAUMATIC STRESS

Steven M. Crimando

The converging trends of increased disaster activity and rapid population growth are forcing professionals across the entire spectrum of emergency management to acknowledge the importance of managing the psychological as well as physical consequences of disasters and terrorism. There are currently more disasters affecting more people than at any other time in recorded history.[1] Terrorism is, of course, a disaster by design, intended to cause the greatest degree of psychological, economic, and social disruption. Whether a disaster is natural, accidental, or terrorist in nature, understanding and addressing the emotional and behavioral impact of an event is critical to effective management.

Across all disciplines involved in emergency management, from homeland security to public health, leaders must be aware of three key behavioral concepts to guide preparedness, response, and recovery:

1. The human response to disasters is "phase specific." People do not react the same way in hour 1 as in day 1, week 1, or year 1. The emotional and behavioral response to disaster continues to change across the timeline. Therefore, all involved in emergency management must anticipate and reflect the phase-specific nature of response in policies, plans, and protocols.
2. The human response to disasters is "hazard specific." People do not react the same way to floods and earthquakes as to mass shootings, disease outbreaks, or radiation emergencies. Planners must therefore be aware of how people actually behave in specific event types.
3. In some disasters the behavioral "footprint" will dwarf the physical or medical impact of the event. But most emergency managers are unprepared to address such situations.

The concept that some disasters are primarily behaviorally based is exemplified by what *Time* magazine has called one of the "worst nuclear disasters" in history.[2]

Local Planning for Terror and Disaster: From Bioterrorism to Earthquakes, First Edition.
Edited by Leonard A. Cole and Nancy D. Connell.
© 2012 John Wiley & Sons, Inc. Published 2012 by John Wiley & Sons, Inc.

GOIÂNIA, BRAZIL

Radiation emergencies, whether accidental or deliberate, pose complex challenges in response and recovery. The radiation emergency in Goiânia, Brazil, in September 1987, serves as a significant reference point in understanding and preparing for both the physiological and psychological challenges associated with these rare but potentially devastating events. Located 1000 miles from Rio de Janeiro and 600 miles from Sao Paulo, Goiânia is the capital of the Brazilian state of Goias. At the time of the incident, Goiânia had approximately one million inhabitants. It is significant to note that the situation occurred in one of the poorest sections of the city, where adult literacy was limited and public health communications were largely undeveloped.

On September 13, 1987, two local men scavenging for scrap metal broke into a deserted clinic in downtown Goiânia and pillaged an orphaned radiation therapy machine and the radioactive source it contained. Sealed inside a containment capsule the source consisted of approximately 20 grams (1375 curies) of Cesium-137 (Ce-137) in the form of cesium chloride salt. Illiterate and unfamiliar with the international radiation symbol, the men did not know that the source was radioactive. Before selling the device to a junkyard, they disassembled the machine and removed the capsule. In the process, they ruptured the container and released the Ce-137 contaminating themselves, nearby family and friends, including young children, and the surrounding environment.[3]

Captivated by the sparkling powder that glowed blue in the dark, one of the men brought the radioactive source to his dining room and called his neighbors and relatives to join in the fun of playing with the substance. People painted themselves with the powder to chase each other around in the night. Within days of the dispersal of the radioactive material, a number of individuals exposed to the substance began to fall ill. On September 29, the junkyard owner's wife filled a bag with the remainder of the powder and traveled by bus to the local hospital. In the process, she contaminated hundreds of people, the bus, and the hospital's facilities.

After initial misdiagnosis as some sort of tropical disease, a local physician recognized the symptoms of acute radiation syndrome. The Comissão Nacional de Energia Nuclear (National Nuclear Energy Commission, or CNEN) was immediately notified, and recognizing the seriousness of the situation, CHEN requested assistance from the International Atomic Energy Agency (IAEA).

Ultimately, this incident resulted in the highest levels of Ce-137 contamination ever clinically recorded, with four deaths and more than 240 people exposed. Several city blocks were demolished and today a protected landfill holds more than 3000 cubic meters of contaminated waste.

The Goiânia accident had an immediate and powerful psychological effect on the population, the victims, and the emergency responders. Despite the challenging technical and medical implications of the event, the psychological response of the affected population proved to be the most daunting. This event yielded a 500:1 ratio of psychological casualties to medical casualties. The Brazilian public health authorities opened the Olympic soccer stadium as a medical screening site and within the first 2 weeks, 112,000 local residents presented for radiation screening.

Most remarkable was that of the first 60,000 screened, 5000 had the signs and symptoms of radiation sickness, although not one had been contaminated. With reddened skin, blisters, vomiting, and diarrhea, the severe psychosomatic symptoms mimicked acute radiation syndrome. The prevalence of these multiple unexplained physical symptoms, accompanied by rumor and misinformation, fueled a surge for emergency health. The local medical and emergency management system was overwhelmed and unable to effectively cope.[4]

Beyond the immediate psychological impact on those believing they had been exposed to a deadly hazard, there was significant damage to the social fabric. Community cohesion quickly devolved into panic and 40,000 people self-evacuated in 4 days. Many burned their clothes and some burned their homes. A few people left the city altogether and tried to change their identities. Airlines refused to sell tickets and hotels declined to rent hotel rooms to anyone thought to be from Goiânia.

Doctors and nurses refused to report for work at the local hospital. Many of the medical staff who did work with the radiation victims developed psychosomatic symptoms and other behavioral disturbances including depression and insomnia. The government's radiation workers were enlisted to care for the sick and injured. There were reports of ambulance drivers abandoning their vehicles with patients onboard because they feared exposure to the patients.

When a 6-year-old girl who played with the radioactive material for several days died, she was buried in a lead coffin sealed in concrete. Her funeral was attended by local residents who, rather than display grief or support her family, stoned the hearse and the grave. They worried that she would continue to pose a radiation risk. The fear and social stigma resulting from the accidental radiation release has persisted in Goiania for a generation.

The failure of government and emergency officials to anticipate and manage the emotional effects of the event complicated the physical response and deepened psychological wounds. Like most emergencies, radiation events prompt behavioral consequences that require proactive management. But research indicates that more than two decades later, those charged with planning for disasters still have a paucity of knowledge in disaster-related behaviors.[5]

In the Goiânia experience, the Brazilian government and local emergency managers with no prior experience in radiation emergencies had no way to fully anticipate the behavioral response to this type of crisis. Among the lessons learned were that emergency management officials benefit from accurate behavioral intelligence and that psychological support mechanisms should be available for victims, the public, and emergency responders. In this and other disasters, it is clear that mental health experts should have been integrated with other emergency response personnel at the onset of response.

BEHAVIORAL CONSIDERATIONS IN EMERGENCY MANAGEMENT

Emergency managers in public sector management agencies and private sector security and safety management programs should incorporate behavioral considerations within their operational procedures. In fact, familiarity with the range of possible behavioral and emotional reactions of both the general population and the responders is essential for policy makers and commanders on the front lines.

Disasters do more than destroy homes, businesses, and public infrastructure; they also leave many individuals and communities traumatized. Disaster-related stress can surface in many forms weeks or months after an event. It may include anger, fatigue, loss of appetite, sleeplessness, nightmares, depression, inability to concentrate, hyperactivity, and/or increased alcohol or drug use. All disasters will result in psychiatric morbidity for a portion of the affected population. Epidemiological research has identified the immediate mental health effects and long-term outcomes of disaster and terrorist incidents.

Dr. Fran Norris reviewed 200 articles published from 1981 through 2001 that addressed the psychosocial impact of disasters on more than 60,000 people.[6] Research in this area suggests that exposure to disasters and terrorism does not necessarily result in post-traumatic stress disorder (PTSD) or other long-term psychological problems. The effects of traumatic events vary greatly among individuals and communities and are influenced by biological and genetic factors, social context, and post-traumatic experiences. Despite the variation in effects, it is a widely held axiom in disaster management that everyone who experiences a disaster, including responders, is affected in some way.

In most of the United States, there are systems within the response network to address the mental health consequences and substance abuse issues that can result from disaster experience. The link between exposure to an event and the onset of mental health and substance abuse problems has led to the incorporation of behavioral health concerns in emergency management and disaster planning initiatives. Federal legislation has required attention to disaster behavioral health since enactment of the Robert T. Stafford Disaster Relief and Emergency Assistance Act more than three decades ago. Specifically, Section 416 authorizes the Federal Emergency Management Agency (FEMA) to fund behavioral health assistance and training activities in areas that have received a presidential disaster declaration. In such instances, a wide range of mental health and substance abuse services for individuals and communities are available upon request by the affected states, with approval of FEMA.[7]

The behavioral health services provided under the FEMA-funded Crisis Counseling Program (CCP) are fundamentally different from traditional mental health services. Crisis counseling helps survivors understand their reactions, improve coping strategies, review their options, and connect with other individuals and agencies that may assist them. While traditional psychotherapeutic approaches seek to change an individual's way of thinking, feeling, and behaving, disaster behavioral health services are intended to help prevent change, and return people to pre-disaster levels of psychological and social functioning.

Equally important, but less prevalent than post-disaster counseling services, is the presence of accurate behavioral information in preparedness, response, and recovery activities. Whereas in most of this chapter the term "behavioral" encompasses mental health and substance abuse issues, in this immediate context the term also describes what people do and do not do in disasters and emergencies. In the wake of the September 11, 2001 terrorist attacks, the emergency management community pushed for "all-hazards" emergency management plans. Intended to help raise the nation's level of readiness for any type of challenge, such plans fail to adequately anticipate human behavior in the face of diverse threats.

In the landmark study, *Redefining Readiness: Terrorism Preparedness through the Eyes of the Public*, Dr. Roz Lasker and associates identified a "fundamental flaw" in emergency preparedness. The flaw is the significant gap between what emergency management officials would have people do in the wake of a terrorist attack or public health emergency and what people would actually do. The study found that "plans to respond to these emergencies *won't* work because people will not react the way planners want them to."[8] Contrary to conventional wisdom, the study found that the public's hesitance to follow official instructions was not due to ignorance or panic, but rather that most people have solid, common-sense reasons for their behavior, even if it is behavior that emergency managers do not anticipate or endorse.

In most jurisdictions, plans have been developed without feedback from the public. Plans typically do not account for all of the risks people would face and do not accurately reflect what people would actually do in different disaster scenarios. The key finding of the Lasker study is "research shows that even if the nation gets all of [the logistics] right, the plans that are being developed now are destined to fail because they are missing an important piece of the puzzle: how the American public would react to these kinds of emergency situations."[9]

THE DISASTER BEHAVIORAL HEALTH SYSTEM

In response to the 9/11 terrorist attacks, President George W. Bush issued a presidential directive that established the National Incident Management System (NIMS) as the single, comprehensive approach required for planning, preparedness, and response for all disaster and emergency response functions across all jurisdiction in the United States.[10] The directive applied to all disaster responders, including behavioral health workers, and facilitated a better integration of disaster mental health and substance abuse services into the overall disaster response system.

Additional guidance in the form of the National Response Framework (NRF) is intended to identify specific authorities and best practices for managing incidents of all types.[11] The NRF and NIMS are the two primary documents that provide overarching guidance in the design of the disaster response system and that apply to disaster behavioral health services.

The NRF defines "the key principles, roles, and structures" that ensure a timely and unified response to disasters and emergencies, and organizes disaster response under 15 Emergency Support Functions (ESFs). The ESF Annex identifies specific federal agencies tasked with leading, coordinating, and supporting the activities of that particular ESF. Most relevant to disaster behavioral health services are ESF #6, #8, and #14.[12]

Because disaster behavioral health services may be provided to support shelter operations and mass care settings, they play an important role in ESF #6: Mass Care, Emergency Assistance, Housing, and Human Services. Disaster behavioral health services are also included in ESF #8: Public Health and Medical Services, where trained personnel may help address the behavioral health needs of both survivors and responders. By providing their usual mental health and substance abuse services, state agencies may also have a response role in ESF #14: Long-Term Community Recovery.[13]

Another essential part of the system is the US Department of Health and Human Services (HHS), and specifically its Substance Abuse and Mental Health Services Administration. This agency and others within HHS work with FEMA to provide technical assistance and training for state and local behavioral health personnel, grant administration, and program oversight. The partnership between FEMA and the relevant HHS groups is identified as the federal Mental Health Authority (MHA). The MHA also stands as an interdisciplinary model for state and county agencies as they strive toward effective emergency management.

An important player in this domain is the Assistant Secretary for Preparedness and Response (ASPR) at Health and Human Services. This individual serves as the HHS Secretary's principal adviser on matters related to bioterrorism and other public health emergencies. This individual also coordinates activities between HHS and relevant federal agencies, and with state and local officials responsible for emergency preparedness and protection of the population in the event of a bio-attack.

Federal agencies recognize only one entity in each state as the designated State Mental Health Authority (SMHA) for the purposes of coordination and communication. While the term "state" is commonly used throughout planning documents, as well as this chapter, it should also be understood to encompass tribal and territorial governments. In most instances, the SMHA is the state's department of mental health, substance abuse, and/or developmental disabilities. Likewise, the federal government identifies and works with each state's emergency management agency. Usually known as the state's Office of Emergency Management (OEM), this entity may be freestanding or part of another body, such as the State Police.

The SMHA and a state's OEM work in partnership across all phases of emergency management (mitigation, preparedness, response, recovery) to ensure that the behavioral health needs of disaster-affected communities are met. Since it is often said that all disasters are local disasters, the state-level agencies, both in behavioral health and emergency management work closely with their county-level counterparts. This helps the state coordinating agencies maintain the NIMS and NRF requirements for a scalable approach to disaster management. Just as at the federal level, county officials tasked with human services and emergency management must also partner to ensure a coordinated response. In many jurisdictions, this approach trickles down to the municipal emergency management system as well.

Disasters touch all parts of life for individuals, families, and communities. To successfully assist those affected by disasters and terrorism, it is critical that responders and response agencies work in close partnerships. Important partners for the behavioral health providers include, but are not limited to, hospital, healthcare, and public health professionals; emergency managers; first responders, including at all levels of law enforcement, fire, and hazmat activity.

Much of this approach is in place in parts of the nation's overall disaster and emergency management structure. But implementation has been uneven—better in some states, tribes, or territories than others. In 2008, the National Biodefense Science Board (NBSB) convened a Disaster Mental Health Subcommittee to make recommendations to address the behavioral health consequences of disasters.[14] In 2010, that subcommittee concluded that "the most pressing and significant problem that hinders integration of disaster mental and behavioral health is the lack of appropriate policy at the highest Federal level. The NBSB cited the lack of any clear statement as to where the authority to devise, formulate, and implement such policy should reside" as the greatest obstacle to effective integration.[15] The subcommittee recommended that a high-level Concept of Operations (CONOPS) be developed to fully integrate disaster behavioral health into "disaster and emergency preparedness, response, and recovery efforts across the Federal enterprise."

To further this effort, toolkits and webinars have been developed for leaders at the state and local levels to provide guidance in developing effective disaster behavioral health plans. To help share resources and lessons learned, as well as to form a unified voice when addressing their federal partners, in 2009 a majority of the states formed the State Disaster Behavioral Health Consortium.[16] Now recognized by their federal counterparts, the consortium serves as a clearinghouse for state level disaster behavioral health coordinators.

Like other critical resources, behavioral health assets can be shared across state lines during a disaster if requested through proper channels. The system of sharing disaster-related resources of all types is the Emergency Management Assistance Compact (EMAC), which is the nation's mutual aid system. Established in 1996, EMAC is a congressionally ratified compact in all 50 states, the District of Columbia, Puerto Rico, Guam, and the US Virgin Islands. It acts as a complement to the federal disaster response system, providing timely and cost-effective relief to states requesting assistance to preserve life, the economy, and the environment. EMAC can be used either in lieu of, or in conjunction with, federal assistance. Through EMAC, disaster behavioral health teams can travel to affected states to supplement their response efforts.

DISASTER BEHAVIORAL HEALTH FUNCTIONS

There have been significant advances in our understanding of traumatic exposure and traumatic stress response in recent years. These advances have helped create the foundation for an evidence-informed approach to early assistance and to resiliency. Well-researched components are now available to support a variety of approaches including:

- Psychological first aid
- Crisis counseling informational briefings
- Crime victims assistance
- Community outreach
- Psychological debriefing
- Psycho-education
- Mental health consultation

The FEMA Crisis Counseling Program (CCP) applies an inclusive approach to recruiting, training, and deploying disaster behavioral health responders. While some disaster relief organizations require that any potential disaster behavioral health responder must be a licensed mental health professional, the CCP model encourages the use of para-professionals and non-professionals as well as those who are licensed.

While it is often rightfully assumed that disaster behavioral health responders will provide crisis or trauma counseling, there is actually a wide range of related services they may undertake. These services may be conceptualized on a continuum from high-intensity/low-volume services to low-intensity/high-volume.

The individual crisis counseling contacts provided through outreach to the affected community (such as a shelter or FEMA Disaster Recovery Center) are an example of a high-intensity/low-volume service. Each contact typically involves an individual counseling session in which the behavioral worker provides emotional support, psycho-education about the normalcy of disaster stress reactions, and ways of coping. Since any one counselor can only reach so many disaster survivors in a day, the volume of their contacts may be relatively low compared to other approaches.

The FEMA Crisis Counseling Program is part of an umbrella of behavioral health approaches required in presidentially declared disasters. The CCP element in the immediate aftermath of a disaster is psychological first aid, a method of early assistance developed by the National Center for Post Traumatic Stress Disorder and other organizations. Psychological first aid includes skills to limit distress and negative health behaviors (e.g., smoking) that can increase fear and anxiety.[17] It is an evidence-informed modular approach to assist children, adults, and families in the immediate aftermath of disaster and terrorism.

Crisis counseling approaches must be adapted to meet the phase-specific reactions of individuals and communities. There are modifications in the approach to crisis counseling across the timeline, including the recent development of a model known as Skills for Psychological Recovery (SPR). SPR is a modification of cognitive behavioral therapy intended for the weeks, months, and even years after a disaster. The approach works by teaching people specific skills that will help them be more resilient. SPR is provided by trained and supervised crisis counselors. The skills taught to disaster survivors include

- problem solving
- planning positive and meaningful activities
- managing disaster stress reactions
- helpful thinking
- building healthy social connections

In presidentially declared disasters, FEMA-funded behavioral health grants to the affected states may be either short term (Immediate Service Program Grants) or longer term (Regular Service Program Grants), ranging from weeks to months in the former instance, and months to years in the latter. Therefore, disaster behavioral health providers should be able to assist survivors with a range of problems across what may be the long timeline of recovery.

Some survivors may still be in distress after psychological first aid, crisis counseling, or SPR. It is also the job of the behavioral health responder to recognize and refer those with lasting or complex behavioral issues to more intensive services, traditionally for mental health or substance abuse treatment. This is indicated when there are the signs of PTSD, anxiety, panic, or depression. A growing body of research supports the use of cognitive behavioral therapy for the treatment of trauma and PTSD, but this is not the sort of intensive clinical work done in the field by disaster behavioral health workers.

Examples of low-intensity/high-volume disaster behavioral health services are media campaigns and public service announcements, the distribution of psycho-educational materials where survivors may gather, and educational presentations in the community or workplace. Through these efforts, trained individuals may reach hundreds or thousands of survivors at one time, though the contacts are not nearly as personal or intense.

There are many barriers to attaining behavioral health support in the wake of a disaster including stigma, damaged roads and vehicles, disruption or lack of financial resources, and insurance issues. It is therefore necessary to reach out to affected communities in a variety of ways. Disaster behavioral health hotlines, web-based and multimedia resources, the availability of bilingual and bicultural counselors—all are necessary to reach broad segments of the population.

ADVANCING THE FIELD OF DISASTER BEHAVIORAL HEALTH

While emergency management and public health initiatives have existed for decades, disaster behavioral health is still an emerging field of management, research, and practice. A system for its delivery has evolved slowly and its integration into emergency management is far from universal.

Efforts to include disaster behavioral health into the national response apparatus are ongoing on many fronts, such as the State Disaster Behavioral Health Consortium and volunteer agencies active in disasters that also provide disaster mental health services. The findings of the NBSB and other groups provide direction for advancing the placement of behavioral health in the overall scheme of disaster management.

An example of the expanding role and future direction of disaster behavioral health is the New Jersey Disaster and Terrorism Branch of the state's Division of Mental Health and Addiction Services RADAR Team. The Rapid Assessment, Deployment and Response team is a small group of highly trained disaster behavioral health workers able to provide on-scene technical assistance to incident commanders, emergency management coordinators, and other leaders. This assistance can help leaders anticipate the likely behavior of a disaster-affected population and mobilize the necessary behavioral health resources, whether in a specific region or statewide. The team is deployed from the state's Regional Operations and Intelligence Center and team members are credentialed by the State Police as essential personnel during times of disaster. They can therefore travel and access disaster scenes more easily.

Another example of the inclusion of disaster behavioral health concerns in national preparedness is the establishment of the Human Factors/Behavioral Sciences Division within the US Department of Homeland Security (DHS). A key objective of the division is to "enhance preparedness and mitigate impacts of catastrophic events by delivering capabilities that incorporate social, psychological and economic aspects of societal resilience."[18] Recognition of the behavioral health impacts of disasters by DHS has helped validate the importance of these issues as an element of national preparedness.

Lastly, innovative technologies may also contribute to addressing a variety of crisis scenarios. New York University's Bioinformatics Group is a collaborative team of experts from the school's Center for Catastrophe Preparedness and Response. They have developed "Plan C"—Planning with Large Agent-Networks against Catastrophes—as a tool for emergency managers, urban planners, public health officials, and emergency responders.[19] By studying the behavioral response to large-scale emergencies, such as a mass food poisoning event in Brazil and the sarin gas attack in Japan, the developers were able to create models and simulations of large, complex emergencies scalable to up to one million casualties. Such a tool can be invaluable to those who must anticipate public behavior in a crisis condition and then effectively manage the crisis.

CONCLUSION

While the role of disaster behavioral health has evolved substantially over the past decades, there remains ample room for growth. The development of initiatives to promote awareness, skills, and tools to meet the psychosocial needs of disaster-affected individuals, families, and communities should remain a priority for leaders in emergency management.

NOTES

1. Mass Casualty Management Systems: Strategies and Guidelines for Building Health Sector Capacity. World Health Organization, 2007.
2. The worst nuclear disasters. *Time Magazine Photo Essay*, March 2011. Available at http://www.time.com/time/photogallery/0,29307,1887705,00.html, accessed October 11, 2011.
3. The Radiological Accident in Goiânia. International Atomic Energy Agency, Vienna, 1988. Available at http://www-pub.iaea.org/mtcd/publications/pdf/pub815_web.pdf, accessed October 13, 2011.

4. Pastel RH. Collective behaviors: Mass panic and outbreaks of multiple unexplained symptoms. Military Med 2001;166:44–46.

5. Lasker RD. *Redefining Readiness: Terrorism Planning Through the Eyes of the Public.* New York: The New York Academy of Medicine; 2004.

6. Norris FH, Friedman MJ, Watson PJ, Byrne CM, Diaz E, Kaniasty K. 60,000 disaster victims speak: Part I. An empirical review of the empirical literature, 1981–2001. Psychiatry 2002;65:207–239.

7. FEMA Fact Sheet: Crisis Counseling Assistance and Training Program, May 2011. Available at http://www.fema.gov/pdf/media/factsheets/2011/dad_crisis_counseling.pdf, accessed October 11, 2011.

8. Ibid., 5.

9. Ibid.

10. Homeland Security Presidential Directive-5. US Department of Homeland Security, February 2003. Available at http://www.dhs.gov/xabout/laws/gc_1214592333605.shtm#1, accessed October 11, 2011.

11. National Response Framework. US Department of Homeland Security, January 2008. Available at http://www.fema.gov/pdf/emergency/nrf/nrf-core.pdf, accessed October 11, 2011.

12. US Department of Health and Human Services, Substance Abuse and Mental Health Services Administration [SAMHSA]. All-hazards disaster mental health and substance abuse preparedness toolkit. 2011, unpublished manuscript.

13. Ibid., 12.

14. National Biodefense Science Board. Disaster mental health recommendations: A report of the Disaster Mental Health Subcommittee of the National Biodefense Science Board, 2008. Available at http://www.phe.gov/Preparedness/legal/boards/nbsb/Documents/nsbs-dmhreport-final.pdf, accessed June 20, 2012.

15. National Biodefense Science Board. Integration of mental and behavioral health in federal disaster preparedness, response, and recovery: Assessment and recommendation, a report of the Disaster Mental Health Subcommittee of the National Biodefense Science Board, 2010. Available at http://www.phe.gov/Preparedness/legal/boards/nbsb/meetings/documents/dmhreport1010.pdf, accessed June 20, 2012.

16. State Disaster Behavioral Health Consortium. Available at http://www .mainedisasterbehavioralhealth.com/Publication9, accessed June 20, 2012.

17. Preparing for the psychological consequences of terrorism: A public health strategy. National Academy of Sciences, Board on Neuroscience and Behavioral Health Institute of Medicine, 2003;108.

18. US Department of Homeland Security, Science and Technology Directorate Human Factors/Behavioral Sciences Division. Available at http://www.dhs.gov/xabout/structure/gc_ 1224537081868.shtm, accessed October 13, 2011.

19. New York University. Planning with large agent-networks against catastrophes (PLAN C). Available at http://www.nyu.edu/ccpr/laser/planc.html, accessed November 2, 2011.

PART IV

SUPPORT AND SECURITY

13

THE ROLE OF THE ON-SCENE BYSTANDER AND SURVIVOR

BRURIA ADINI

DIZENGOFF SHOPPING MALL

On March 4, 1996, hundreds of children with their parents were out celebrating Purim, a Jewish holiday that commemorates the deliverance of the Jewish people in the ancient Persian Empire from destruction. Many were in costumes, similar to those seen in other societies on Halloween. At approximately 3:50 P.M., a suicide bomber exploded himself while walking on the sidewalk adjacent to the Dizengoff Center Mall, a central metropolitan attraction.[1] He had strapped to his body 20 kilograms of explosives including a large number of nails to maximize the damage. His intention to enter the mall itself was deterred as he noticed a security officer at the entrance, so he detonated the explosives while standing next to a group of people at a stoplight. Thirteen were killed and 125 injured.[2]

Dozens of bystanders and survivors rushed to the scene, parallel to the activation of professional first responders. Many of them displayed symptoms of stress and confusion. Some were crying, screaming, running back and forth looking for their loved ones, or speaking loudly to one another. Others started to offer help to the victims, inquiring if the emergency medical services (EMS) and the police force had been alerted, if there was a need to evacuate casualties to hospitals, or if they could assist in directing traffic away from the area.

Within minutes, the street was jammed with vehicles, first responders, and bystanders, packing the intersections leading to the mall; security and evacuation vehicles found it difficult to enter or leave the scene. The general impression was one of chaos.

Manner of Communication and Interaction with Other Responder Groups

The survivors and bystanders interacted with three main responder groups: the police, EMS, and admitting hospitals. As there were no designated communication means (cell phones

Local Planning for Terror and Disaster: From Bioterrorism to Earthquakes, First Edition.
Edited by Leonard A. Cole and Nancy D. Connell.
© 2012 John Wiley & Sons, Inc. Published 2012 by John Wiley & Sons, Inc.

were less common then), the typical channel for communicating with other responders was face-to-face. Several official first responders tried to convince the bystanders to leave the scene. The bystanders were instructed that for their own well-being, as well as that of the casualties, it would be best if the professional teams could work independently and uninterrupted. But most bystanders stayed in place. They kept asking for the opportunity to participate in the rescue operations and provide help to the victims.

Successes and Failures of the Response

The futile attempts to compel bystanders or survivors to leave the scene impeded rapid control of the situation. Crucial resources were invested in ineffective argument and communication between police officers and bystanders. But some of the EMS personnel approached the situation differently. They requested bystanders and survivors who were not injured to assist as stretcher-bearers and evacuate casualties to the ambulances. They also directed individuals to stand at different intersections adjacent to the site and advise drivers not to enter the mall area. The majority of bystanders adhered to the requests and assisted in facilitating the management of the mass casualty incident (MCI). This chapter provides an analysis of how bystanders react to disasters and terror events, and how bystanders might be successfully integrated into the emergency response.

Utilization of Bystanders and Survivors

Chaos and confusion are dominant characteristics of terror events and disasters,[3,4] caused in part by bystanders themselves.[5,6] Many people rush to the scene to find out what occurred, even when requested to stay away from the area. Once there, most people appear to respond spontaneously and generously to victims of a disaster and are willing to provide assistance to ease suffering.[7] An effective mechanism to utilize bystanders, then, is to direct the actions of those individuals who arrive at the scene that offer assistance. During disasters and terror events, people look for leadership[8] and most often obey orders by authorities. Crowd control and integrating the activities of the bystanders and survivors can be a fruitful component of the emergency response and positively impact the results of the disaster.[8] Nevertheless, for this collaboration to occur, prior planning must take place. The roles of all those involved must be clearly defined so that the execution of these plans will succeed.

OVERVIEW

The primary causes of chaos at the scene of terror events and disasters are

1. obstruction of accessibility by bystanders drawn to the scene out of curiosity and/or wish to offer assistance
2. disruption of reliable information flow and creation of rumors
3. loud and unfamiliar noises
4. confusion among both the professional responders and the general public
5. feelings of helplessness, resulting from the horrific repercussions of the disaster

If not dealt with effectively, bystanders and survivors of the event contribute significantly to the chaos and have the potential to magnify its consequences. Many bystanders think that their active assistance following an emergency will be productive, but studies have shown that their actual behavior differs from their preconceptions.[9] Lack of leadership on-site may worsen the situation, whereas clear and defined expectations accompanied by direct orders allocating the bystanders and survivors with specific and defined tasks will facilitate management of the situation. Bystanders who witness a terrorist event can be expected to participate to facilitate an optimal response capability.[10]

Some studies report that bystanders and survivors tend to help only in situations that they perceive as beneficial and low risk.[11] Other studies support the view that bystanders are an important resource for developing an effective emergency response.[12] They can perform numerous tasks on-site, as illustrated by the report of a survivor of a suicide bombing event on September 19, 2002, in Tel Aviv. The bomber detonated a bomb in Bus Number 4, killing the driver and four passengers, and wounding an additional 60. When the smoke cleared, Heyn, 87 years old, saw a heart, still beating, lying on the sidewalk amidst shattered glass.

> There was no body—just a heart beating. I didn't think such a thing could happen. But I saw it and others saw it. Glass shards fell on me. I looked to see what had happened to me, but I was not hurt. I saw a man walking covered with blood. I took off my shirt and used it as a bandage to help him.[13]

This example is just one among many that describes how the reaction of a bystander or survivor could contribute to the response model for a terrorist or disaster event.

First Responders

First responders include emergency public safety authorities, fire brigades, ambulance services, police and law enforcement agencies, emergency response teams, and related personnel, agencies, and authorities.[14–16] They are responsible for the protection and preservation of life, property, and environment.[17] Yet bystanders and survivors should also be considered as potential first responders in view of their availability, accessibility, and capacity to provide immediate assistance to victims and responders following a terror event. Since they are already at the scene, they can offer assistance before disaster officials arrive.[18] Their integration as an important component of the response model can contribute to saving lives and improving the management of the event.

Functions that Bystanders and Survivors Can Perform

Reporting of the Event   Activation of official first responders is dependent on their receiving notification of the occurrence of a terrorist event.[19] The first report provided to the operation center will initiate the deployment of vital resources (manpower, evacuation vehicles, equipment and other necessary infrastructure). The accuracy of information that is relayed (what happened, the type of event, the scope of resources that are needed, and best access and egress routes) will determine the appropriateness of the deployment.[20] The first notification is most often received from bystanders or survivors of an emergency event. If handled appropriately, bystanders and survivors can access and disseminate these crucial details.

Reconnaissance   Terrorist events, in particular explosions, are often characterized by the distribution of casualties over a wide area. First responders must search for all victims, and there is an urgent need to locate and identify the surviving casualties.[21] Certified first responders may need assistance in searching for and locating casualties that are not immediately visible. Bystanders and survivors can assist in this activity and report their findings, thus alerting first responders and directing them to additional casualties.

Assistance in the Triage of Casualties   Following life-support interventions by the EMS personnel, non-professional assistance may be needed due to a lack of sufficient certified personnel on-scene. Bystanders and survivors can be activated to assist in numerous medical treatments, such as maintaining an open airway or applying direct pressure to control bleeding. Clear instructions must be provided by the EMS staff to instruct the bystander how to perform these tasks. Bystanders who have received first-aid training can have a significant impact on the outcome of the injuries.[22] However, when acting independently, bystanders are limited in their capacities to provide medical assistance[23] and they tend not to use their skills and proficiencies.[24] They may at times be unable to recognize emergencies and delay crucial medical treatments.[25] Therefore, their successful deployment requires careful direction and supervision.

Caring for the Walking Wounded   Lack of sufficient resources is one of the major complexities of an MCI due to terrorism. Improving management of non-urgent casualties reduces the burden on the first responders and improves resource utilization.[26] After triage determination, a bystander or survivor can assist in tending to patients with minor injuries until medical teams or evacuation facilities are available.[4,27] The participation of bystanders in this role can ensure that the walking wounded receive attention while also preventing the rush of non-urgent casualties into ambulances, reserving them for the more severely injured.

Providing attention to the walking casualties allows the professional teams to focus on triage and treat the severe or moderate casualties.[28] In addition, the mildly injured may sometimes be insistent on attention and if left unattended, may obstruct the work of the EMS paramedics. By providing this assistance, bystanders contribute to the ability of medical teams to work according to MCI protocols and thus contribute toward saving lives.

Assisting in Traffic Control   In many terrorist events, roads and intersections leading to and from the site of the event very quickly become jammed with traffic, blocking access to the scene and evacuation efforts.[6,29,30] Effectively utilizing bystanders can help avoid this phenomenon. Police or other security officers can direct bystanders or survivors to divert traffic away from strategic points such as intersections or roads leading directly to the site of the event. Such a deployment can achieve a dual purpose: (1) the bystanders assist in decreasing obstruction of traffic by redirecting cars to other destinations, and (2) as a result of performing this task, the bystanders themselves move further away from the scene and thus minimize their interference with emergency response procedures.[6]

Strengthening Security   In the last decade, several terrorist events involved a second explosive device which was detonated several minutes after the initial explosion, with the aim of injuring or killing the first responders that have already arrived on-site.[31] This threat led to recognition of the grave need to seal off disaster scenes. Bystanders and survivors can then be deployed to direct any passersby or drivers to alternate destinations. Bystanders can also alert security officers to suspicious individuals and/or findings and assist in collecting criminal or other types of evidence.

Evacuation of Casualties to Ambulances   Shortage of manpower, including stretcher-bearers, is characteristic in emergencies, both on-site and in hospitals.[5] This activity requires little training, and bystanders and survivors can fulfill this mission with dual benefits: the professional first responders will be freed to provide lifesaving procedures to the casualties and the bystanders will be satisfied that they can contribute to managing the event. Recruiting passersby as stretcher-bearers has been shown to be successful in both prehospital (carrying the casualty from the place of injury to the ambulance) and hospital (transporting casualties from the ambulance to the emergency department and within the different treatment sites) settings.

Transporting Casualties to Medical Facilities   MCIs often suffer from an imbalance between urgent needs and available resources.[32] In routine incidents, such as road accidents, some casualties are evacuated to medical facilities by non-EMS vehicles.[26,33,34] It has been reported that though the majority of casualties are evacuated during MCIs by ambulances, many of them are transferred by non-EMS vehicles.[35] This phenomenon is even more frequent during terror events due to a delay in arrival of sufficient number of ambulances compared to the immediate availability of private cars of bystanders who are willing to evacuate casualties to a hospital.[36]

Collection of Body Parts   Terrorist events, especially in the case of massive explosions, may result in the dispersal of body parts across a wide area.[37] Many religions consider the gathering of all body parts for burial as a holy act. This activity continues after evacuation of all casualties to acute care hospitals and may be a prolonged process. This mission must be conducted carefully, with extreme sensitivity and under supervision of official agencies, such as the police force or other law enforcement agencies.

Collection of Personal Belongings to Facilitate Identification of Victims   To facilitate identification of casualties involved in terror events, it is of utmost importance that all personal belongings of the victims be collected and handed over to the police force or other agencies responsible for the identification process.[29,38] This task too must be performed with great sensitivity and supervised to prevent any misconduct or mishandling of the belongings.

Accompanying Relatives to Forensic Institute   Terror events almost always occur without prior expectation or notification. Family members of the victims are suddenly required to participate in the process of locating or identifying the remains of loved ones. In many countries, the relatives of suspected victims must go to the Forensic Institute or other professional agency responsible for the identification process.[38] Formal representatives of welfare or medical organizations might be unavailable to assist relatives in this tragic process. Bystanders and survivors can be recruited to accompany and support worried family members. This activity must also be conducted with sensitivity and in coordination with the formal agencies, police and/or medical officials.

Locating Family Members Separated during the Event   On-site confusion and chaos and evacuation of casualties to hospitals may lead to separation of victims from the same family. Some may be transferred to different hospitals.[39] Assistance in reuniting relatives or providing information of their whereabouts should take place as soon as possible, especially when children are involved. Bystanders can be used for this activity under the supervision of social workers from the local municipality or the acute care hospitals.

Assisting in Operating Information Centers   Immediately upon the occurrence of a terrorist event, there is a crucial need to open information centers to relay information to worried relatives.[40] Lack of available information may lead to vain searches by numerous family members and friends. Information regarding the victims must be collected by hospital administrations or the local municipality, who are then responsible for its dissemination to the general public. In fact, bystanders can also be deployed in these information centers to assist in calming worried relatives, supplying beverages and food, and providing a "warm touch." They can also be recruited to accompany family members during the process of waiting to receive information.

Assisting the Media   Reporting of information from the scene of a terrorist event is one of the dominant aims of the media. In the immediate wake of an MCI, covert competition begins among the different media providers to provide coverage from the scene of the event. Managing the media response appropriately is vital to protect the control and command among first responders, to maintain communication structures and to mitigate stress among both providers and the public.[41] Bystanders and survivors can provide first-hand impressions of what had happened including personal stories that are often welcomed by both reporters and the public. Most individuals own sophisticated cell phones or other advanced electronic devices, and they can provide the media with verbal reports and photographs or video coverage of the event. These materials prove useful to the media and are needed also for investigating and documenting what occurred on-site.

Translators   Migrants and visitors from other countries who are not fluent in the native language(s) may find a disaster event even more frightening and confusing than the resident population. Bystanders speaking different languages can serve as translators during the mass casualty event and can assist communication between the professional first responders and the victims.[42]

Blood Donations   Availability of blood units for the injured is a crucial component of an emergency response.[43] An appropriate and practical method for blood donations needs to be created for operation both in routine times and in emergencies. National blood banks frequently suffer from shortage of crucial blood units, despite various programs that are initiated in order to raise awareness and compliance of the public to donate blood. In times of emergencies, especially those caused by intentional man-made MCIs, the public tends to cooperate and is willing to donate blood.[44,45] If the blood bank has made prior preparations and is capable of rapidly organizing a massive blood donation operation, mobile units can be brought to the scene of the terror event to request bystanders and survivors to donate blood.

ASSESSMENTS

Familiarity with the Required Functions

In most societies, bystanders and survivors are not yet recognized as an integral component of the response model for terrorist events or other disasters. This lack of recognition may result in a less effective management of the event since, as discussed, bystanders may obstruct the certified first responders from acting according to professional guidelines. This is likely due to citizens' lack of familiarity with triage principles, leading to misinterpretation of some actions. For example, they might not understand a decision to withhold treatment for extremely severe casualties.

If not integrated appropriately, bystanders and survivors may squander limited resources of the first responders, obstruct evacuation routes, and even make themselves more vulnerable for secondary damage.[46] However, when utilized appropriately, a bystander's intervention may help reduce the number of casualties.[47]

Readiness to Perform These Functions

The readiness of bystanders and survivors to assist in various types of emergency scenarios has not been assessed under non-emergency conditions. Observations of real-life emergencies suggest that bystanders are more likely to intervene in emergency situations than in non-emergency scenarios.[48] In the case of a clear emergency, anonymity influences the decision regarding one's own obligation to intervene; when it is unclear whether a situation is an emergency, anonymity of the bystanders delays their decision-making process as to whether to help.[49] It has also been demonstrated that the likelihood of giving assistance tends to decrease when the number of bystanders increases.[50] Thus, there is a need for professional first responders to convey to bystanders an understanding of their value and contribution to the outcomes of a terrorist event during an emergency.

Identification of Other Disciplines with Which Bystanders Might Interact

One of the basic components of on-site management during an MCI is the control and command, also known as incident command.[51–53] Regardless of the agency that is responsible for this activity there has to be a well-defined authority that dictates to all involved agencies the boundaries of their on-site involvement.[8] This is relevant also with regard to the bystanders and survivors of the event. The responsible control and command agency is the most important authority with which the bystanders interact, as it has the capacity to allow (or disallow) the bystanders to be integrated into the response mechanism.

It is the decision of the scene commander whether to deploy bystanders and survivors.

The major on-site interaction of the bystanders is likely to be with the EMS personnel, as they can benefit the most from the assistance of non-professional individuals for immediate actions. Hospitals can also benefit from the integration of bystanders in non-complex activities such as performing administrative tasks, acting as stretcher-bearers, and accompanying family members in the information centers. Similar activities can be carried out by bystanders in information centers that are deployed by local municipalities.

Additional disciplines with which the bystanders and survivors interact are volunteer organizations that operate at the event and are willing to recruit them as immediate reinforcement.

PLANS AND EXERCISES THAT INCLUDE BYSTANDER INTERACTION

The involvement of bystanders in institutional, local, regional, or national exercises is a complex process. Many exercises are conducted with first responders to test and expand their knowledge and capabilities for managing terrorist events or other types of emergencies.[54,55] Yet the willingness of bystanders to take part in these drills may be lacking in routine times.[56]

While their willingness is very high during real-life events, in everyday life most civilians show little incentive to participate in preplanned training or exercise activities.[57] It is also not feasible for the exercise planners and executers to include a population over whom they have no authority. Therefore, bystanders can rarely be recruited for training in an exercise.

Nevertheless, as illustrated here, the integration of bystanders in the planning process could be extremely helpful. The roles that might be allocated to them, the mechanism for their integration, and their interaction with the different disciplines must be prepared for prior to the event itself. Such guidelines can and should be implemented in advance within the Standard Operating Procedures of the various first responders, who will be the major stakeholders to benefit from the recruitment of bystanders.

COMMUNICATIONS AND OTHER CAPABILITIES

MCIs are situations in which multi-agency organizations and multidisciplinary teams must collaborate, working under chaotic and highly dynamic environments.[58] A diverse work-force from numerous cross-jurisdictional agencies, some of which have not worked together prior to the event, have to coordinate to perform their tasks.[58] As the environment is chaotic, the basic perception of many representatives of the first responder agencies, such as police and EMS personnel, is that the scene of a terrorist event should be handled only by the professional organizations. Often the prevailing attitude is that non-certified individuals obstruct rather than facilitate the management of the event, and that all efforts should be made to distance them from the scene. Friction often develops between bystanders and first responders during terrorist events or other types of MCIs.

As mentioned earlier, attempts to coordinate the deployment of the bystanders and survivors are even more difficult due to lack of familiarity with communication structures and the command hierarchy that characterize the activity of formal first responders.

Another limitation is that there are as yet no formal mechanisms for communication between bystanders and first responders. Most individuals have immediate access to cell phones or other types of communication channels, but there is no mechanism to link these individuals once an event has taken place. Nonetheless, it might be feasible to gather this information during the event itself and transmit it to the relevant dispatch centers. Many mobile phones are equipped with two-way video capacities that enable the callers to both hear and see each other. This real-time visual information could further help senior medical staff provide bystanders with supervision and guidance.[59]

CONCLUSION

As proposed here, bystanders and survivors should be an integral component of response efforts for terrorist events or other types of disasters. Lack of effective deployment of bystanders can impede effective management of the event. Integration of bystanders in coordination and collaboration with the certified first responders will contribute to saving lives, enhance societal cohesiveness, and raise the resilience of the community.

However, bystanders and survivors are likely to facilitate the response only if their potential roles are recognized prior to the event and are implemented as part of prior preparedness planning. Certified first responders, such as police or EMS personnel, as well as the interface agencies, such as hospitals and local councils, must be trained in how to integrate bystanders in the emergency response. The potential contribution of bystanders and survivors should be publicized in order to encourage their inclusion in the planning and preparedness process for terrorist and other disaster events.

NOTES

1. Weiss M. The body of the nation: Terrorism and the embodiment of nationalism in contemporary Israel. Anthropol Quart 2002;75(1):37–62.

2. Cohen-Almagor R. Media coverage of terror: Troubling episodes and suggested guidelines. Canadian J Commun 2005;30(3):383–409.

3. Frykberg ER. Principles of mass casualty management following terrorist disasters. Ann Surg 2004;239(3):319–321.

4. Lowe CG. Pediatric prehospital medicine in mass casualty incidents. J Trauma 2009;67(2): S161–S167.

5. Mayo A, Kluger Y. Terrorist bombing. World J Emerg Surg 2006; 1: 33.

6. Singer AJ, Singer AH, Halperin P, et al. Medical lessons from terror attacks in Israel. J Emerg Med 2007;32(1):87–92.

7. Quarantelli EL. Conventional beliefs and counterintuitive realities. Social Res 2008; 75(3):873–904.

8. Pinkert M, Bloch Y, Schwartz D, et al. Leadership as a component of crowd control in a hospital dealing with a mass-casualty incident: Lessons learned from the October 2000 riots in Nazareth. Prehosp Disaster Med 2007;22(6):522–526.

9. Demirovic J, Li YP. Bystanders of out-of-hospital sudden heart attack: Knowledge and behaviors among older African Americans. Am J Geriatr Cardiol 2005;14(4):171–175.

10. White RD. Optimal access to- and response by-public and voluntary services, including the role of bystanders and family members, in cardiopulmonary resuscitation. New Horiz 1997;5(2):153–157.

11. Avdeyeva TV, Burgetova K, Welch ID. To help or not to help? Factors that determined helping responses to Katrina victims. Anal Soc Iss Pub Pol 2006;6(1):159–173.

12. Taylor PA, Levine M, Best R. Intra-group regulation of violence: Bystanders and the 'de'-escalation of violence. IACM 21st Annual Conference Paper, 2008. Available at http://ssrn.com/abstract=1298601, accessed June 20, 2012.

13. Report of a Civilian Coalition in Israel. *Burning Flowers, Burning Dreams: Consequences of Suicide Bombings on Civilians in Israel, 2000–2005*. Chapter 8. Israel, 2009. Available at http://suicidalterror.com/references.htm, accessed June 20, 2012.

14. Benedek DM, Fullerton C, Ursano RJ. First responders: Mental health consequences of natural and human-made disasters for public health and public safety workers. Annu Rev Pub Health 2007;28:55–68.

15. Elmqvist C, Brunt D, Fridlund B, et al. Being first on the scene of an accident—Experiences of 'doing' prehospital emergency care. Scand J Caring Sci 2010;24(2):266–273.

16. Reynolds CA, Wagner SL. Stress and first responders: The need for a multidimensional approach to stress management. Int J Dis Manag 2008;2(2):27–36.

17. VanDevanter N, Leviss P, Abramson D, et al. Emergency response and public health in Hurricane Katrina: What does it mean to be a public health emergency responder? J Pub Health Manag Pract 2010;16(6):E16–E25.

18. Helsloot I, Ruitenberg A. Citizen response to disasters: A survey of literature and some practical implications. J Contingencies Crisis Manag 2004;12(3):98–111.

19. Schafer WA, Carroll JM, Haynes SR, et al. Emergency management planning as collaborative community work. J Homeland Sec Emerg Manag 2008;5(1):Article 10.

20. Shapira SC, Cole L. Medical management of suicide terrorism. In: Falk O, Morgenstern H, editors. *Suicide Terror: Understanding and Confronting the Threat*. Hoboken, NJ: John Wiley & Sons; 2009. pp 381–395.

21. Byers M, Russell M, Lockey DJ. Clinical care in the "Hot Zone." Emerg Med J 2008; 25(2):108–112.

22. Tomruk O, Soysal S, Gunay T, et al. First aid: Level of knowledge of relatives and bystanders in emergency situations. Adv Ther 2007;24(4):691–699.

23. Breckwoldt J, Schloesser S, Arntz HR. Perceptions of collapse and assessment of cardiac arrest by bystanders of out-of-hospital cardiac arrest (OOHCA). Resuscitation 2009;80(10):1108–1113.

24. Swor R, Khan I, Domeier R, et al. CPR training and CPR performance: Do CPR-trained bystanders perform CPR? Acad Emerg Med 2006;13(6):596–601.

25. Messmer PR, Jones SG. Saving lives. An innovative approach for teaching CPR. Nurs Health Care Perspect 1998;19(3):108–110.

26. Bloch YH, Leiba A, Veaacnin N, et al. Managing mild casualties in mass-casualty incidents: Lessons learned from an aborted terrorist attack. Prehosp Disaster Med 2007;22(3):181–185.

27. Sharma BR. Development of pre-hospital trauma-care system—An overview. Injury 2005;36(5):579–587.

28. Aylwin CJ, Konig TC, Brennan NW, et al. Reduction in critical mortality in urban mass casualty incidents: analysis of triage, surge, and resource use after the London bombings on July 7, 2005. Lancet 2006;368(9554):2219–2225.

29. Scanlon J. Sampling an unknown universe: Problems of researching mass casualty incidents (a history of ECRU's field research). Stat Med 2007;26(8):1812–1823.

30. Almgody G, Bala M, Rivkind A. The approach to suicide bombing attacks: Changing concepts. Eur J Trauma Emerg Surg 2007;33(6):641–647.

31. Goh SH. Bomb blast mass casualty incidents: Initial triage and management of injuries. Singapore Med J 2009;50(1):101–106.

32. Baker MS. Creating order from chaos. Part I: Triage, initial care, and tactical considerations in mass casualty and disaster response. Mil Med 2007;172(3):232–236.

33. Armstrong JH, Frykberg ER, Burris DG. Toward a national standard in primary mass casualty triage. Disaster Med Public Health Preparedness 2008;2:S8–S10.

34. Raiter Y, Farfel A, Lehavi O, et al. Mass casualty incident management, triage, injury distribution of casualties and rate of arrival of casualties at the hospitals: Lessons from a suicide bomber attack in downtown Tel Aviv. Emerg Med J 2008;25(4):225–229.

35. de Ceballos JP, Turégano-Fuentes F, Perez-Diaz D. 11 March 2004: The terrorist bomb explosions in Madrid, Spain—An analysis of the logistics, injuries sustained and clinical management of casualties treated at the closest hospital. Crit Care 2005;9(1):104–111.

36. Pinkert M, Lehavi O, Goren OB, et al. Primary triage, evacuation priorities, and rapid primary distribution between adjacent hospitals—Lessons learned from a suicide bomber attack in downtown Tel-Aviv. Prehosp Disaster Med 2008;23(4):337–341.

37. Hiss J, Kahana T. Forensic investigation of suicide bombings. In: Shapira SC, Hammond JS, Cole LA, editors. *Essentials of Terror Medicine*. New York: Springer; 2009. pp 393–403.

38. Morgan OW, Sribanditmongkol P, Perera C, et al. Mass fatality management following the South Asian tsunami disaster: Case studies in Thailand, Indonesia, and Sri Lanka. PLoS Med 2006;3(6):e195.

39. Welzel TB, Koenig KL, Bey K, Visser E. Effect of hospital staff surge capacity on preparedness for a conventional mass casualty event. West J Emerg Med 2010;11(2):189–196.

40. Adini B, Peleg K, Cohen R, et al. A national system for disseminating information on victims during mass casualty incidents. Disasters 2010;34(2):542–551.

41. Nocera A. Disasters, the media and doctors. Med J Aust 2000;172(3):137–139.

42. Steffen SL, Fothergill A. 9/11 volunteerism: A pathway to personal healing and community engagement. Social Sci J 2009;46(1):29–46.

43. Yi M. Investigation into the practice of ensuring a steady blood supply for medical rescue during the Wenchuan Earthquake. J Evid Based Med 2009;2(3):158–163.

44. Turégano-Fuentes F, Diaz DP, Sanz-Sánchez M, et al. Overall assessment of the response to terrorist bombings in trains, Madrid, 11 March 2004. Eur J Trauma Emerg Surg 2008;34(5):433–441.

45. Liu J, Huang Y, Wang J, et al. Impact of the May 12, 2008, earthquake on blood donations across five Chinese blood centers. Transfusion 2010;50(9):1972–1979.

46. Dean R, Mulligan J. Management of home emergencies. Nurs Stand 2009;24(6):35–42.

47. Harrison M. Bombers and bystanders in suicide attacks in Israel, 2000 to 2003. Studies Conflict Terrorism 2006;29(2):187–206.

48. Shotland RL, Huston TL. Emergencies: What are they and do they influence bystanders to intervene? J Pers Soc Psychol 1979;37(10):1822–1834.

49. Schwartz SH, Gottlieb A. Bystander anonymity and reactions to emergencies. J Pers Soc Psychol 1980;39(3):418–430.

50. Jo HH, Jung WS, Moon HT. Rescue model for the bystanders' intervention in emergencies. Europhys Lett 2006;73(2):306–312.

51. Adini B, Laor D, Cohen R, et al. The five commandments for preparing the Israeli healthcare system for emergencies. Harefuah 2010;149(7):445–450, 480 [Article in Hebrew].

52. Cryer HG, Hiatt JR, Eckstein M, et al. Improved trauma system multicasualty incident response: Comparison of two train crash disasters. J Trauma 2010;68(4):783–789.

53. McCann DGC, Cordi HP. Developing international standards for disaster preparedness and response: How do we get there? World Med Health Policy 2011;3(1):Article 5.

54. Hughes LE, Derrickson S, Dominguez B, et al. Economical emergency response and lower profile states: A behavioral health reference for the rest of us. Psychol Trauma 2010;2(2):102–108.

55. Descatha A, Loeb T, Dolveck F, et al. Use of tabletop exercise in industrial training disaster. J Occup Environ Med 2009;51(9):990–991.

56. Wiese CHR, Wilke H, Bahr J, et al. Practical examination of bystanders performing basic life support in Germany: A prospective manikin study. BMC Emerg Med 2008;8:14.

57. Wilkerson W, Avstreih D, Gruppen L, et al. Using immersive simulation for training first responders for mass casualty incidents. Acad Emerg Med 2008;15(11):1152–1159.

58. Culley JM, Effken JA. Development and validation of a mass casualty conceptual model. J Nurs Scholarsh 2010;42(1):66–75.

59. Bolle SR, Johnsen E, Gilbert M. Video calls for dispatcher-assisted cardiopulmonary resuscitation can improve the confidence of lay rescuers—Surveys after simulated cardiac arrest. J Telemed Telecare 2011;17(2):88–92.

14

THE ROLE OF THE TRAINED VOLUNTEER

Brenda D. Phillips, Njoki Mwarumba, and Debra Wagner

HURRICANES AND THEIR AFTERMATH

Between 2005 and 2008, Gulf Coast hurricanes tested the readiness and willingness of all responder groups including those of trained volunteers. In August 2005, Hurricane Katrina generated significant numbers of volunteers as did Hurricane Rita in the following month and Hurricanes Ike and Gustav in 2008. One group of trained volunteer responders, the Medical Reserve Corps (MRC), was the product of a relatively recent initiative. Established in 2002, the MRC is one of several organizations that comprise the Citizen Corps. Most of them were formed after 9/11. The MRC began as a test concept, but by 2007 it had grown to 695 units with more than 127,000 volunteers throughout the United States, Puerto Rico, Guam, the US Virgin Islands, and Palau.[1]

Composed of already-trained medical personnel, the MRC has the ability to support local, transitory, or emergent public health needs including in evacuation, triage, shelters, and medical care. Hurricanes Katrina and Rita, which struck Louisiana and some other states with unusual force, tasked several MRC units to assist. Trained volunteers heeded the call. They were especially effective with Rita, moving in advance of the event to support evacuees along the coast, further inland, and in host states as far away as Nebraska. One of their principal tasks was to provide support for higher-risk populations that were subject to transfer trauma—those moved from hospital rooms, assisted living facilities, or private homes to a range of locations: university arenas, airport facilities, public campgrounds, extended care facilities.

Taking care of hundreds of frail patients, some very ill, required an extensive turnout. Physicians, nurses, emergency medical technicians, pharmacists, and dentists all responded through their local units. Approximately 6000 MRC volunteers assisted with both Katrina and Rita, with about 100 units serving outside of their own jurisdictions.[2] A uniformed

Local Planning for Terror and Disaster: From Bioterrorism to Earthquakes, First Edition.
Edited by Leonard A. Cole and Nancy D. Connell.
© 2012 John Wiley & Sons, Inc. Published 2012 by John Wiley & Sons, Inc.

Texas Medical Rangers MRC supported emergency medical care at the Dallas Convention Center, and Dallas Union Rita created a Disaster Hospital site in Tyler, Texas.[3]

Many lessons were learned from the 2005 experiences. Evidence from the 2008 hurricane season shows that the efforts of trained MRC volunteers resulted in considerable financial savings. MRC responders to Hurricane Gustav in Lake Charles, Louisiana, helped with triage, evacuation, sanitation, telephone hotlines, and at shelters. Approximately 100 MRC volunteers donated more than 2000 hours of time with a savings estimated at \$76,252.[4] MRC units helping the Kentucky Department of Health sent 47 volunteers to staff a medical-needs shelter for Gustav evacuees, and a total of \$19,400 was saved in labor costs.[5] In Oklahoma, a veterinary MRC established a shelter for evacuees' pets, donating \$4444 in time and service.

Among the most important lessons from these experiences is that as disaster and emergency agencies become strained, trained volunteers can be immensely helpful.

Trained volunteers can be important assets before, during, and after a crisis. Because they bring knowledge and skills to all phases of disaster, they can enhance preparedness initiatives, support responders, help during recovery, and incorporate mitigation measures to reduce losses. Organizations that support trained volunteers offer a valuable resource by insuring that volunteers arrive on a scene fully aware of the risks associated with volunteering: from dealing with toxic contamination to picking up debris to safe reconstruction practices. Organizations often facilitate the arrival, dispersal, and support of volunteers including transportation, housing, feeding, safety, supervision, and recognition. The result is the availability of valuable savings and support for emergency professionals.

Without the participation of dedicated, trained, and willing volunteers like those seen in the MRC, people and animals awaiting the impact of disaster would suffer. Volunteers must be recruited, trained, and motivated to stay involved and form a base in which disaster-resilient communities can develop.

OVERVIEW OF TRAINED DISASTER VOLUNTEERISM

Jurisdictions and their resources vary widely, from rural and isolated areas dependent on part-time professionals to large cities with cadres of paid employees. Rural areas may even need to rely on volunteers to staff their fire and emergency management offices. Because of the limited number of professionals experienced in disaster impacts, most first responders and emergency managers rely on volunteers. There is good news concerning the potential turnout of trained volunteers. Most societies host populations socialized to be altruistic toward those in need. Such pro-social mindsets inspire and motivate volunteer turnout for all kinds of locations and events.[6]

Professionals at terrorist or disaster events will experience varying levels of involvement with volunteers. Consider, for example, the case of the May 3, 1999 tornado outbreak in Oklahoma that claimed 41 lives and destroyed hundreds of houses across the state. In the small community of Bridge Creek, those who responded included school officials, a small volunteer Red Cross unit, and a handful of police, fire, and emergency medical technicians. Training paid off, as volunteers rendered desperately needed aid not only during the night of the tornado but for days afterward. Working together, they stepped out of familiar roles. With only basic disaster training, they saved many of their neighbors' lives, cared for the deceased, organized massive amounts of donations, and hosted a federal disaster recovery center.

Emergency and disaster professionals need to tap into such social capital. Volunteer centers in some communities may be a good source for securing trained volunteers, especially in locations that have previously experienced a disaster. Volunteer centers can also be employed to organize rapid low-skill training, such as might be needed for sandbagging. But the best kind of trained volunteer comes in with advance certification and experience. Toward that end, federal agencies have supported development and training of Citizen Corps which, as noted, comprises the Medical Response Corps and other volunteer groups. In addition, many disaster-specific organizations provide trained or supervised volunteers. For example, the National Voluntary Organizations Active in Disaster (NVOAD) links disaster-stricken communities to more than 100 disaster-experienced organizations. The majority of NVOAD affiliates come from the faith-based community such as Presbyterian Disaster Assistance and Lutheran Disaster Services. These organizations alone can provide thousands of volunteers during relief and recovery activities.

Volunteer Functions in a Terrorist or Disaster Event

A wide array of trained volunteers can be found in most communities, though with varying abilities. Should a local community lack volunteer resources, they can be developed (see Resources Section at the end of this chapter). This section reviews how trained volunteers can help through all phases of disaster: preparedness, response, recovery, and mitigation. Citizen Corps groups may, for example, help in preparedness activities by educating the public about disaster threats. MRC volunteers may be extremely valuable during the early stages of response, especially among medically fragile populations. At the outset of a major disaster, the Federal Emergency Management Agency (FEMA) usually opens a joint field office where the Voluntary Agency Liaison (VAL) coordinates with volunteer organizations helpful during recovery.[7]

After Hurricane Katrina, FEMA opened a joint field office in Baton Rouge. FEMA-VAL convened the first post-disaster voluntary organization recovery meeting. Organizations present then fanned out to identify and address unmet needs. Affiliated volunteers provided dental, medical, spiritual, and psychological care; interorganizational coordination for massive donation storage and distribution; repairs and reconstruction; childcare and therapeutic children's camps; substitute clergy for churches, mosques, and synagogues; and mitigation measures, such as elevating homes and installing hurricane clamps on roofs.[8] Without trained volunteers (including those who supervise spontaneous unannounced volunteers), a full response and recovery effort would have been impossible. Organizations that produce trained volunteers may be divided into three categories: disaster organizations, community-based organizations, and faith-based organizations.

Disaster Organizations that Train Volunteers

The American Red Cross trains volunteers in such general functions as first aid, mental health, and shelter management. The organization also trains about 11 million people annually in first aid and cardiopulmonary resuscitation (CPR) alone with a large base of 158,000 volunteer instructors (see www.redcross.org). Additional programs support medical personnel for agent-specific events such as chemical, biological, and radiological releases. Both in-person and web-based training is available to volunteers. With hundreds of chapters throughout the United States, chapter-affiliated Disaster Action Teams represent crucial and well-prepared volunteer teams. An organization that expands and contracts, the Red Cross can dispatch significant numbers of volunteers and paid staff to disaster locations. When a major event occurs, the Red Cross can staff hundreds of shelters with trained volunteers who may come from all parts of the nation. Their depth of experience generates a ready core of volunteers. Indeed, the Red Cross responds to 70,000 disasters a year including an average of 150 house fires daily. The experience from these small-scale events means that Red Cross volunteers maintain a knowledge and skill base that could also help with large-scale disasters.

Salvation Army volunteers, often called the "army behind the Army," also provide extensive and experienced support. Known for food and beverage canteens, the Salvation Army moves into damaged areas to support first responders and affected residents. The organization's volunteers also comfort those in need of emotional and spiritual care. The Indian Ocean tsunami of 2004 challenged Salvation Army units worldwide to respond. Their work began at the moment of impact with staff and volunteers pulling people from the water, providing medical care, and helping those searching for relatives. They continued their work through offering food supplements and relief supplies. Later, the Salvation Army assisted with rebuilding by funding home reconstruction, small business recovery, and school redevelopment (for more information, visit www.salvationarmyusa.org).

Community-Based Organizations that Train Volunteers

In 2002, the US Congress began to fund community-based disaster response teams. Given that most people affected by disaster will be aided first by neighbors and fellow citizens, the Citizen Corps offers a crucial asset close to those at risk. Organized under the broad concept of a nationwide Citizen Corps, these groups include Community Emergency Response Teams, Medical Reserve Corps, Volunteers in Police Service, Neighborhood Watch, and Fire Corps. This broad cadre of local volunteers is geared toward responding to a variety of hazards and giving support to professionals with emergency response functions.[9]

Community Emergency Response Teams   Community Emergency Response Teams (CERTs) build grass-roots "first responder" capacities in local communities. Their origin is traced back to 1985 when CERTs first appeared in Los Angeles.[10] In 1993, FEMA began offering "train-the-trainer" courses at a national level, consequently changing CERT by institutionalizing, standardizing, and promoting it.[11] Today the organization comprises more than 1100 listed community groups.

CERT focuses on educating communities about disaster preparedness and response activities that may affect their specific location, an opportunity that resonates well with community-minded citizens.[12] Training includes providing basic disaster response information including first aid, search and rescue, team organization, and disaster medical operations (http://www.citizencorps.gov/cert/faq.shtm). Upon completion, CERT members work as emergency responders alongside fellow trained volunteers, or perform supporting tasks for professional responders. CERT training can be used in a variety of settings to instruct locally available volunteers.

As an illustration, teenagers represent an available base, although young volunteers are more likely to come from institutions that support service learning.[13] Students in Winter Springs High School in Seminole County, Florida, became involved in such an effort. Their CERT training included survival skills that culminated in a tabletop exercise, which earned high school credit for each participant. This school-based program succeeded in enlarging the collective pool of trained volunteers. Furthermore, it stimulated peer interest with potential for attracting more young volunteers. The program's success has further spurred the growth of CERTs in communities and colleges throughout the nation.

Medical Reserve Corps   Trained medical volunteers are a critical part of medical and public health response. The events of 9/11 and the subsequent anthrax attacks demonstrated the need to develop a cadre of medical and nonmedical volunteers willing to respond when public health and medical resources threaten to become overwhelmed. Since 2002, when the MRC was created as a Citizen Corps partner program, the MRC has become an integral part of emergency planning at the local, state, and federal levels.

The MRC program provides a framework for recruiting, screening, and training volunteer and performance management.[14] Sponsored by the Office of the Surgeon General, the Division of the Civilian Volunteer Medical Reserve Corps now hosts 202,700 volunteers who are members of 973 teams nationwide. The division also serves as a clearinghouse for information about resources and best practices. Regional consultants support state and local MRC coordinators, and local units then develop according to the needs of their particular areas.

MRC units are community-based assets in which volunteers can identify their skills and interests and receive necessary training to become effective responders. Core training competencies include personal and community preparedness, incident command, and psychological first aid. Members participate in emergency response drills, community events, and public health initiatives with a focus on improving community resiliency.

Volunteers in Police Service   Following the 9/11 terrorist attacks, US law enforcement personnel experienced a growth in demand for their services. In addition, increased citizen interest in volunteer law enforcement resulted in creation of the USA Freedom Corps to incorporate volunteers into service (http://www.policevolunteers.org/about/). The Volunteers in Police Service (VIPS), a program developed out of this initiative, is managed by the International Association of Chiefs of Police, in partnership with other agencies including the US Department of Justice (www.policevolunteers.org). In most communities, however, the creation of VIPS teams remains under local control.

Standardized training modules, academy training, and certification are offered to qualified volunteers, who are then attached to local agencies. VIPS volunteers may become involved in a variety of activities, such as issuing parking citations, supporting search and rescue, processing paperwork, policing special events, staffing phone banks, and maintaining law enforcement vehicles.[15]

Neighborhood Watch   The Neighborhood Watch program, established in 1972, currently operates under the USA on Watch program. Neighborhood Watch trains residents to observe residential areas to prevent crime and increase area safety. Their awareness and social networks can be tapped into for emergency preparedness activities as well. A key premise of the Neighborhood Watch program is that by strengthening informal networks, local residents can enhance each other's quality of life by reducing crime (http://www.usaonwatch.org/resource/neighborhood_watch_toolkit.aspx?).

Fire Corps   Inaugurated in 2004 under Citizen Corps, Fire Corps volunteers are trained to assist with nonemergency/nonoperational tasks in fire and emergency management. At the national level, Fire Corps is managed by the National Volunteer Fire Council and other agencies associated with the Fire Corps National Advisory Committee. Similar to

VIPS, Fire Corps volunteers supplement resource-strained and under-budgeted fire and emergency service departments. Locally based training is delivered through traditional in-class methods as well as online media. Funding comes largely from grants secured through local Citizen Corps Councils and federal agencies including the Department of Homeland Security.

Fire Corps volunteers may help with administrative duties, apparatus and facility maintenance, fund raising, rehabilitation services, public education, and fire hydrant maintenance (http://www.firecorps.org/about-us). Beyond assisting fire departments, Fire Corps benefits the broader community by exposing citizens to the value of first responders, what they require to support fire response, and how to prepare at home.

Faith-Based Organizations that Train Volunteers

Many faiths have some type of disaster organization affiliated with a particular denomination. Faith-based organizations (FBOs) such as the United Methodist Committee on Relief, Presbyterian Disaster Assistance, or Lutheran Disaster Response offer large numbers of potential volunteers. These people typically arrive with varying levels of training. Still, FBOs usually offer all the volunteers on-site training or supervision for disaster relief and recovery work. Some FBOs have generated programs that train and credential volunteers in a particular specialty.

When disaster strikes, a range of needs will emerge, not all of which can be met by government agencies, first responders, or emergency managers. Tornadoes that ravage communities, for example, will generate large amounts of debris. Volunteers may be tasked to rapidly collect and discard materials to help jump-start recovery.

Some FBOs have organized internally to expedite relief and recovery work. Mennonite Disaster Service, for instance, sends experienced disaster relief teams into the field to help with chain saw work or cleaning up muck. The Mennonite service also deploys site investigators to determine what needs exist. Short-term volunteers may serve for 1 or 2 weeks. If an area meets with their protocol, well-trained long-term volunteers might set up project sites and volunteer housing. After Hurricane Katrina, about 5000 volunteers continued to provide annual service to coastal areas for several years. Their work included the rebuilding of homes and restoration of faith in humanity. Such tangible and intangible benefits of trained volunteer teams helped heal communities both physically and spiritually.

Disasters associated with terrorism offer particular challenges. The emotional effects of a deliberately caused incident are often more intense than one of accidental or natural origin. Organizations had to reconfigure traditional responses, for example, to the Oklahoma City bombing and to the 9/11 attacks. In these cases, FBOs paid for medical care and funerals, assisted with transportation needs, and supported traumatized responders. The Lutheran Disaster Response Chaplain Network trains and ordains pastors for such work. Designed as a means to guide people through crisis, the Chaplain Network provides emotional care and spiritual care in the face of devastating psychological injury (for more information, see http://www.ldr.org/care/chaplain-network.html).

Disasters can also disrupt childcare arrangements, preventing caretakers from returning to work. The Church of the Brethren Disaster Ministries trains and certifies childcare workers for such needs. First, the volunteers receive background checks. They then are taught how to offer therapeutic play, emotional support, and safety after a disaster. The church also equips them with kits designed for children's post-disaster needs.[16]

Many individuals and households require assistance beyond what insurance, personal savings, and government aid provide. Typically, local long-term recovery committees identify and help their neighbors with unmet needs. Within those committees, case managers (voluntary though sometimes paid) present client cases and solicit aid from those present (groups like the Mennonite Disaster Service, Buddhist Tzu Chi Society, Catholic Charities and Nazarene Disaster Response). In recent years, the United Methodist Committee on Relief has worked with FEMA to design and train local workers on disaster-appropriate case management procedures. They trained hundreds of case managers to assist Texans affected by Hurricanes Katrina in 2005 and Ike in 2008. Most recently, the Methodist group has generated publicly available webinar training on case management (see Resources Section).

FBOs provide a structure through which volunteers can leverage their collective energies, resulting in trained and supervised volunteers for a range of emergency activities. Furthermore, faith groups typically operate more flexibly than bureaucratically oriented government agencies. Such flexibility means that they can adapt to unanticipated circumstances and reconsider their approaches, particularly in the face of new and emerging disasters. As an illustration, FBOs have recently focused on pandemic planning in an effort to scope out an appropriate support role for their volunteers. Church World Service offers web-based modules with preparedness information on H1N1 and other viruses. FBOs disseminate this information throughout their communities. As the public continues to face new threats, the faith community will likely further adapt. A massive electrical grid failure due to a space or weather event is one such possibility. In such a case, faith groups could provide essential help to those with medical conditions and disability needs.[17]

KEEPING THE TRAINED VOLUNTEER INTERESTED

The dynamic base of volunteerism means that familiarity with functions and roles can vary and change. Disasters typically drive interest in becoming a trained volunteer. Events like 9/11 or the tornado that struck Joplin, Missouri, in 2011 inspire people to learn first aid, search and rescue, and other skills. Over time, though, trained volunteers may lose skills unless they engage in periodic exercises. They may also lose interest in volunteering if their skills go unused. This can be especially problematic since major events may be infrequent and lengthy periods may pass without the need for trained volunteers.

Still, organizational abilities to train and maintain a ready cadre of volunteers are essential for response capacity. Established disaster organizations, such as the American Red Cross or the Citizen Corps, are well positioned to generate trained volunteers. These types of organizations deserve wide support from professional managers and responders as well as elected officials. Currently, federal funds are available to the states for Citizen Corps groups. States then determine how to use these dollars, whether for training, conferences, or small grant programs. At the local level, the character of the community often influences volunteer interest. Communities with a strong tradition of civic activity are more likely to generate and maintain a strong volunteer body.

Equally important is the nature of the connection between voluntary organizations and the emergency response and management sectors. Emergency professionals who rely exclusively on paid staff fail to provide opportunities for volunteers to learn, and this could lead to staffing problems when a disaster occurs. Thus, emergency professionals would be wise to involve volunteers in lesser emergency activities, which would also help keep their interest. Emergency management agencies could involve CERTs in a range of activities:

- distributing preparedness information
- conducting mitigation assessments
- analyzing local populations for social vulnerabilities
- assisting residents to develop personal and pet preparedness kits
- creating plans and exercises for disaster response

Non-disaster events can also be used to maintain interest levels by tasking volunteers with various responsibilities such as traffic management at sporting events, festivals, parades, and public health outreach. Trained volunteers can be sent to daycare centers, schools, senior centers and into universities to assess safety procedures, list recommendations, and distribute information. Keeping volunteers engaged is critical.

A good example of interorganizational connection occurred in Watsonville, California, after the 1989 Loma Prieta earthquake. Since Latinos comprised 60% of the population, outreach was attempted in Spanish and through culturally sensitive approaches. After encountering some problems with communication, the local Red Cross hired a tri-lingual, tri-cultural specialist well-versed in Latino, Native American, and mainstream American approaches. The Red Cross then connected with a local organization, Salud Para La Gente (a Latino health organization) and offered training to its board, staff, and membership, who then returned the favor. As a consequence, the community emerged with a committed interorganizational set of trained volunteers ready to assist in the next disaster.

READINESS

The state of readiness varies by location and resources. Americans are willing to step up in time of disaster, but few seem willing to undergo the necessary training in advance. Still, training and credentialing volunteers are essential to attaining the best possible outcomes.[18] With volunteers prepared for specific tasks, emergency professionals can feel more secure about who enters a scene and what they can do. Trained volunteers would also be prone to fewer injuries (such as from chain saws, downed power lines, or toxic contaminants). These volunteers would also be better prepared to face emotionally wrenching situations and more likely to recover from psychological trauma.[19]

Training and placing a volunteer in a functional unit like CERT is most important to attract and keep recruits. Response to recruitment is often dependent on the organization and professionalism of the recruiting units. Well-organized MRCs, for example, attract trained people willing to dedicate their time.[20] Similar results have been found for FBOs which draw volunteers when they provide good structure and support.

ADDITIONAL PARTNERS FOR TRAINED VOLUNTEER INVOLVEMENT

Groups dedicated to developing disaster volunteers are not the only source of potential volunteer help. Community-based and service organizations also attract volunteers with varying levels of training and expertise.[21] When a disaster occurs, such civic-minded volunteers can often be counted on for support. As one illustration, after Hurricane Katrina, advocates for people with disabilities emerged from the broader community. Some advocates moved into support positions with emergency professionals and solved problems with sheltering, communications, and temporary housing. With insights provided by those closest to highly vulnerable groups, disaster professionals can better organize relief for these populations. This situation occurred in New Orleans when faith-based groups assisted with evacuation and recovery of Vietnamese Americans.[22]

By identifying the range of social services and relevant organizations in a community, a disaster professional can seek assistance from a larger pool of possible volunteers. Communities affected by disaster can then expect to move more quickly from response to recovery. Emergency managers typically reduce their presence as the recovery unfolds and local leaders (elected officials, appointees, volunteers) organize long-term recovery committees. By pre-identifying potential partners, training them on disaster issues, and incorporating them into relief and recovery plans, the emergency professional increases the community's ability to organize post-disaster recovery.

Nor should emergency professionals overlook individuals who are unaffiliated with existing organizations. Too often, people at risk in disasters (particularly seniors and people with disabilities) are presumed dependent and unable to make a contribution. Yet their experiences can also yield valuable information about how best to reduce the impacts of disasters through preparedness, response, and recovery initiatives. Savvy emergency managers will seek avenues to reach those typically at high risk in a disaster and provide them with training—through schools for the deaf, senior citizen centers, social clubs, and places of worship. Many such individuals could then be integrated into disaster-specific organizations. Officials in Denver (in Colorado) and Pittsylvania County (in West Virginia) did exactly that by offering CERT classes to residents who were deaf. By providing training with the help of interpreters, they increased the core of volunteers who could help both the general public and those with similar sensory challenges.

Colleges and universities also provide learning opportunities for engagement in disaster work through internships and research projects. When the Great Hanshin-Awaji earthquake struck Kobe, Japan, in 1995, nursing students helped senior citizens who were experiencing isolation, alcohol abuse, and depression that could lead to suicide.[23] After the 2011 tornado outbreak in Alabama, students poured into areas that needed help with medical care, debris removal, childcare, and needs assessment.

Similar efforts took place in Vermont after Hurricane Irene, as students moved into communities to help with debris removal. By integrating disaster work into "service learning" courses, faculty can help prepare students for response efforts. Many disciplines are potential partners for such service including social work, nursing, emergency management, fire protection, psychology, dentistry, veterinary care, environmental studies, architecture, construction, and more. Student labor coupled with faculty guidance has thus far been a largely untapped source for trained and supervised volunteers.[24]

PLANS AND EXERCISES THAT FOSTER INTERACTION

At the national level, the Top Officials (TOPOFF) exercise program has involved up to 20,000 participants in each exercise, including volunteers and their related organizations. Offered every 2 years, TOPOFF scenarios could apply to multiple areas across the United States. Many of the exercises include mock attacks with biological, chemical, or radiological agents. Such a model could be emulated at state and local levels, and several emergency management professionals have been doing just that. The level at which this occurs varies by jurisdiction. Oklahoma, for example, held a statewide, full-scale exercise in April 2011. In conjunction with the exercise, about 65 MRC volunteers were trained and utilized as evaluators.

Oklahoma's MRC also activates volunteers during each exercise and launches quarterly "call down" drills of their entire database. (A call down is a pre-arranged procedure for phoning information to relevant parties.) Working across organizations and recognizing local hazards, the MRC involved volunteers in Operation Firestorm in the fall of 2011. The full-scale exercise at the Ardmore Airpark included a simulated explosion aboard a commercial airliner with 100 fatalities and 65 injuries.

COORDINATION AND COMMUNICATIONS WITH TRAINED VOLUNTEERS

During a response, it is critical to bring volunteer liaisons into the emergency operations center. Typically, this may occur under a FEMA program known as Emergency Service Function (ESF) #6, which involves mass care, emergency assistance, housing, and human services. Other functional areas may benefit from such a liaison as well, such as Fire Corps—ESF #4 covers firefighting. Most importantly, such coordination should be part of pre-event exercises and not initiated at the time of an event. In sum, proper interorganizational coordination requires participation of voluntary organization leaders when developing pre-event response and recovery plans.

The issue of interoperable communication systems is also relevant to trained volunteers. The radio frequencies and jargon used by teams that provide communications equipment to volunteers may vary. Thus, cross-team training is advisable as is the notion of cultural interoperability. Volunteers may operate from a different mindset than emergency professionals. Understanding the professionals' approaches can make communication and coordination between the two groups more effective. To illustrate, the term "incident command" though commonly understood by disaster responders may not be familiar in some voluntary organizations. Pre-event, interorganizational training, and practice are crucial to avoiding confusion on such issues.

Most groups that support trained volunteers use e-mail lists, websites, and other electronic means to stay in touch. After the tornado outbreak across Alabama in 2011, the state used its Facebook page to alert volunteers to points of assembly. Alabama also used its 2-1-1 phone-in system to register volunteers and then activated volunteer reception centers across the state. The spontaneous and trained volunteers who turned out conducted extensive help.

But dependence of these networks can sometimes be problematic. During Hurricane Isabel in 2003, Virginia CERTs were in the midst of e-mailing and phoning when the area's power failed. Afterward, two-way radios were made available to various CERT groups in case of future power loss.[25] Still, the vulnerability persists as Facebook, Twitter, and other electronically based social media become more common.[26]

CONCLUSION

It is clear that people in all parts of the world try to help in time of disaster. They give blood, donate money, and provide comfort to soften the impact of hurricanes, earthquakes, floods, or terror attacks. Trained volunteers are a further expression of the human impulse to help. They occupy a niche that can substantially mitigate the effects of disaster, for which they deserve appreciation. Yet many agencies and organizations that offer training programs remain under-resourced and under-funded. These groups warrant support from communities and emergency professionals, who stand to gain from their neighbors' willingness to provide help and enhance community relationships. Such volunteerism often spills over to nonemergency activities as well by inspiring people toward greater civic involvement. Recruiting and training volunteers represents a major step that any organization can take toward building a more disaster-resilient community.

RESOURCES

- Information on Citizen Corps can be found at www.citizencorps.gov
- United Methodist Committee on Relief (UMCOR) Case Management training is available on the Church World Services Emergency Response Program website as a 90-minute webinar, see http://www.cwserp.org/
- The National Voluntary Organizations Active in Disaster (NVOAD) offers a list of disaster-affiliated organizations that link into varying levels of trained volunteers. Many assist in the recovery and reconstruction phases of a disaster.

NOTES

1. Hoard M, Middleton G. Medical Reserve Corps: Lessons learned in supporting community health and emergency response. J Bus Contin Emer Plan 2008;2:172–178.

2. Ibid.

3. Greenstone J. The Texas Medical Rangers in the military response of the uniformed Medical Reserve Corps: To Hurricane Katrina and Hurricane Rita 2005. Available at http://www.dtic.mil/cgi-bin/GetTRDoc?Location=U2doc=GetTRDoc.pdfAD=ADA499311, accessed October 2, 2011.

4. Schaffzin S. Impact of the Medical Reserve Corps on hurricane response efforts. Disaster Med Public Health Prep 2009;3:126–127.

5. Ibid.

6. Rodriguez H, Trainor J, Quarantelli EL. Rising to the challenges of a catastrophe: The emergent and prosocial behavior following Hurricane Katrina. Annals Am Acad Political Social Sci 2009; 604:82–101. Lowe S, Fothergill A. A need to help: Emergent volunteer behavior after September 11. In: Monday J, editor. *Beyond September 11th: An Account of Post-Disaster Research*. Boulder, CO: Natural Hazards Research and Applications Information Center; 2003. pp 293–314.

7. FEMA. The role of voluntary agencies in emergency management. Independent Study 288. 1999. Available at http://www.fema.gov, accessed September 30, 2011.

8. Phillips B, Jenkins P. The roles of faith-based organizations after Hurricane Katrina. In: Kilmer K, Gil-Rivas V, Tedeschi R, Calhoun L, editors. *Meeting the Needs of Children, Families, and Communities Post-Disaster: Lessons Learned from Hurricane Katrina and Its Aftermath*. Washington, DC: American Psychological Association; 2009.

9. Drabczyk A. Ready, set, go: Recruitment, training, coordination, and retention values for all-hazard Partnerships. J Homeland Sec Emerg Manag. 2003;4:Article 12.

10. Schulz C, Koenig K. A medical disaster response to reduce immediate mortality after an earthquake. New Engl J Med 1996;334:438–444.

11. Ibid.

12. Flint C, Stevenson J. Building community disaster preparedness with volunteers: Community emergency response teams in Illinois. Nat Hazards Rev 2010;11:118–124.

13. Sundeen R, Raskoff S. Volunteering among teenagers in the United States. Nonprofit and Voluntary Sector Quarterly 1994;23:383–403.

14. Connors T, editor. *The Volunteer Management Handbook*. New York: John Wiley & Sons; 1995.

15. Hart A. Sonoma County law enforcement shows up in droves to honor volunteers. 2011. Available at http://rohnertpark.patch.com/articles/sonoma-county-law-enforcement-shows-up-in-droves-to-honor-volunteers, accessed October 2, 2011.

16. Peek L, Sutton J, Gump J. Caring for children in the aftermath of disaster: The Church of the Brethren Children's Disaster Services Program. Children Youth Environ 2008;18:408–421.

17. Office of the Federal Coordinator of Meteorology. 2010 Space Weather Enterprise Forum, Summary Report. Available at http://www.ofcm.gov/swef/2010/SWEF_2010_Summary_Report_%28Final%29.pdf, accessed March 1, 2011.

18. Britton NR. Permanent disaster volunteers: Where do they fit?. Nonprofit Voluntary Sector Quarterly 1991;20:395–415. Clizbe JA. Challenges in managing volunteers during bioterrorism response. Biosecur Bioterror 2004;2:294–300. Paton D. Training disaster workers: Promoting wellbeing and operational effectiveness. Disaster Preven Manag 1996;5:11–18. Paton D. Disaster relief work: An assessment of training effectiveness. J Trauma Stress 1994;7:275–288.

19. Dyregrov A, Kristoffersen JI, Gjestad R. Voluntary and professional disaster workers: Similarities and differences in reactions. J Trauma Stress 1996;9:541–555. Thoits PA, Hewitt LN. Volunteer work and well-being. J Health Soc Behav 2001;42:115–131.

20. Qureshi K, Gershon RM, Conde F. Factors that influence Medical Reserve Corps recruitment. Prehosp Disaster Med 2008;23:S27–S34.

21. National Council on Disability. *Effective Emergency Management: Making Improvements for Communities and People with Disabilities*. Washington DC: NCD; 2009.

22. Airriess C, Li W, Leong KJ, et al. Church-based social capital, and geographical scale: Katrina evacuation, relocation and recovery in a New Orleans Vietnamese American community. Geoforum 2008;39:1333–1346.

23. Kako M, Ikeda S. Volunteer experiences in community housing during the Great Hanshin-Awaji earthquake, Japan. Nurs Health Sci 2009;11:357–359.

24. Kushma J, Phillips B. Service learning. For the FEMA Higher Education Project, 2002. Available at http://training.fema.gov/EMIWeb/edu/sl_em.asp, accessed June 20, 2012.

25. Franke ME, Simpson DM. Community response to Hurricane Isabel: An examination of CERT organizations in Virginia. Quick Response Research Report 170, Natural Hazards Center, University of Colorado, Boulder, 2004.

26. Hovey WL. Examining the role of social media in organization–volunteer relationships. Public Relations J 2010; 4:1–23.

15

BIOTERRORISM, BIOSECURITY, AND THE LABORATORY

Nancy D. Connell and James Netterwald

DIAGNOSIS ANTHRAX

Sunday, September 29, 2001: Robert and Maureen Stevens were driving in North Carolina with their daughter, Casey, who was living in Charlotte.[1] The three had spent the previous day hiking on the nearby Appalachian Trail. Suddenly, Stevens, a 63-year-old photojournalist, was overcome with flulike symptoms: high fever, chills, and weakness. He felt better after resting overnight, and he and his wife drove back home to Florida the next day, Monday. Shortly after midnight, Stevens was weak and dizzy; his wife drove him to the emergency room at the John F. Kennedy Medical Center in Atlantis. His temperature was extremely high and he was delirious and disoriented. After an initial diagnosis of meningitis he was admitted and placed on antibiotics. Hours later he suffered a seizure and was intubated, and his cerebrospinal fluid was tested for infection.

On the morning of Tuesday, October 2, Dr. Larry Bush, an infectious disease specialist, was asked to look at a sample from Stevens' spinal tap. Under the microscope, the bacteria showed all the hallmarks of the genus *Bacillus*—dark blue staining, boxcar shape, large size. Bush reviewed his thoughts:

> The organism was a bacillus, as evident by its Gram stain and its shape . . . When you think of the common bacilli that can cause somebody to be ill, there is one called Bacillus cereus, which you can see with traumatized patients, or with immune-compromised patients. Another is Bacillus subtilis, which, again, we occasionally see in the bloodstream . . . this patient had no reason to have any bacillus as far as exposures or trauma were concerned. He had not been an ill person and he had no immune system defects.[2]

Dr. Bush turned to the colleague who had prepared the sample and asked her opinion. Kandy Thompson, an experienced medical technologist, suggested that the bacteria might

Local Planning for Terror and Disaster: From Bioterrorism to Earthquakes, First Edition.
Edited by Leonard A. Cole and Nancy D. Connell.
© 2012 John Wiley & Sons, Inc. Published 2012 by John Wiley & Sons, Inc.

be *Clostridium*, the genus that causes tetanus and botulism. But Bush continued his line of reasoning; the bacteria were larger than clostridia, could even be *Bacillus anthracis*.

Although there was some increased attention to biological weapons during the 1990s, at the time of the anthrax attacks, most physicians were unfamiliar with the agents commonly considered potential biological weapons, such as *Yersinia pestis* (plague) or *Variola virus* (smallpox). A strong base of knowledge for diagnosis and treatment of plague and smallpox were available, but only 18 cases of inhalational anthrax had been identified in the United States during the twentieth century. The most recent fatal case had been a California home craftsman, who succumbed in 1976 after exposure to bacteria isolated from imported yarn.[3] Bush's presumptive diagnosis was astute and remarkable, and "took matters across a divide with immense and frightening implications."[4]

The next step was to confirm the diagnosis by sending a sample to the Integrated Regional Laboratories in Fort Lauderdale. The results of two more tests were available about 6 hours later: both motility and hemolysis were consistent with the organism's being *B. anthracis*. Bush sent the sample overnight to the state health laboratory in Jacksonville. By mid-afternoon on Wednesday, October 3, two of three additional tests had been performed: the bacteria had tested positive for the anthrax capsule, a thick protective coat, and by 10 P.M. that evening, for the presence of a specific polysaccharide in its cell wall. The definitive test—sensitivity to a bacteriophage (virus)—would take another overnight incubation. Ironically, Phil Lee, the microbiologist who was overseeing the testing, had just returned from a workshop at the Centers for Disease Control and Prevention (CDC) to learn how to perform these tests. Early on October 4, Bob Stephens was definitively diagnosed as infected with *B. anthracis*. He died the next day at 3:55 P.M.

BIOLOGICAL ATTACK: THE CONTEXT

The release of anthrax spores in 2001 through the US postal system was arguably the most high-impact bioterrorism event in modern history. The American people, still reeling from the 9/11 attacks, were dealt a second blow in a largely unfamiliar format: biological terrorism. The decade preceding 2001 saw a dramatic increase in attention to the problem of biological weapons. Concerns were heightened with the discovery in the 1990s of secret Soviet and Iraqi bioweapons programs. Official concerns remain unabated, as evidenced in a 2011 statement by Alexander Garza, Chief Medical Officer of the Department of Homeland Security:

> The threat of an attack using a biological agent is real and requires that we remain vigilant. A wide area attack using aerosolized Bacillus anthracis, the bacteria that causes anthrax, is one of the most serious mass casualty biological threats facing the U.S. An anthrax attack could potentially encompass hundreds of square miles; expose hundreds of thousands of people, and cause illness, death, fear, societal disruption and economic damage. If untreated, the disease is nearly 100 percent fatal, which means that those exposed must receive life-saving medical countermeasures as soon as possible.[5]

Despite acknowledgment of this reality, many experts feel that we are still unprepared for a biological attack. In 2010, the bipartisan Commission on the Prevention of Mass Destruction Proliferation and Terrorism gave the Obama administration a grade of "F" for bioweapons preparedness. Led by former Senators Bob Graham and Jim Talent, the commission explained:

> The absence of a comprehensive U.S. capability to rapidly recognize, respond to and recover from a disease based attack is the most significant failure identified in the final 'report card' . . . The national capacity should include the ability to provide information to authorities and the general public; adequate supplies of medical countermeasures and program of rapid distribution of those medicines; systems for isolating and treating the sick; and environmental cleanup systems for materials such as anthrax.[6]

Meanwhile, biomedical technology continues to advance rapidly. Increased understanding of biological agents (bacteria, viruses) and of the host (immunology, neurology, physiology) has led to the potential for exploitation of medical knowledge for nefarious purposes.[7] This concern was recently brought to international attention by experimental developments in the field of influenza pathogenesis.[8]

This chapter focuses on the roles that scientists and healthcare workers played in the aftermath of the 2001 anthrax attacks, and how they might contribute to building a web of preparedness for a biological event, whether of natural or intentional origin.

BIOLOGICAL DEFENSE PROGRAMS IN THE UNITED STATES IN 2001

The US Army Medical Research Institute for Infectious Diseases (USAMRIID) was established in 1969 at Fort Detrick, in Frederick, Maryland. The laboratory was built at the site of the country's former offensive bioweapons program, which was terminated in 1969 by President Richard Nixon. In 1972, the United States signed on to the Biological Weapons Convention (BWC), the first arms control program to prohibit the development or stockpiling of an entire class of weapons. USAMRIID became the flagship laboratory of the country's new defensive biomedical research program. Suspicions of a continuing Soviet biological weapons (BW) program sustained US research and development of vaccines and therapeutics for agents that cause diseases such as anthrax and plague.

In 1979, an outbreak of anthrax in the Soviet city of Sverdlovsk raised questions about possible illegal BW activity in that area. Eventually, it was learned that the outbreak had resulted from an accidental release of anthrax spores from a secret military production facility.[9] The accident provided impetus to USAMRIID's efforts during the 1980s to optimize the existing anthrax vaccine. These efforts were further bolstered when Iraq, a country with a suspected BW arsenal, invaded Kuwait in 1990, thereby precipitating US troop engagement during the 1991 Gulf War Iraq. (As later confirmed, Iraq had developed an arsenal of anthrax and other biological agents, though it did not use them in the conflict.)

The first World Trade tower attacks in 1993 and the Oklahoma City bombing and Tokyo subway chemical attacks in 1995 underscored the risks of terrorism faced by US citizens. The Clinton administration then began to focus more on civilian defense,[10] and funds for biodefense became available to several federal agencies.[11]

Awareness in particular of the threat of weapons of mass destruction (WMD) increased, resulting in new funding streams for WMD preparedness. For example, the US Department of Defense established the National Domestic Preparedness Program, which provided support for such preparedness to the country's 120 largest cities.[12] It was in this context, amid increasing awareness and activity regarding preparedness, that the anthrax mailings took place.

DISTINCTIVE NATURE OF A BIOLOGICAL EVENT

Biological events lack the single-episode quality of a chemical, nuclear, or conventional-explosive disaster. Biological attacks are characterized by an incremental recognition of patterns of morbidity and/or mortality, and eventual realization that an outbreak has occurred. The slower timeline of an actual event may suggest ample time to respond, yet experience shows that careful preparation for disease outbreaks is necessary to protect lives. Furthermore, as with other kinds of disasters, planning, training, and exercising is crucial. A global surveillance mechanism and a strong public health infrastructure are key to identifying and containing outbreaks.

In the United States, public health investigators fall within a section of the CDC called the Epidemiological Investigation Service (EIS). The EIS trains these healthcare professionals in the investigatory techniques of outbreak analysis—many of these trained officers played crucial roles during the anthrax attacks.

Like other terrorist events, the anthrax mailings had a criminal component. Both the FBI and the investigatory arm of the US Postal Service launched a massive investigation. According to the Department of Justice website, the Amerithrax Task Force consisted of roughly 25–30 full-time investigators from the FBI, the US Postal Inspection Service, and other law enforcement agencies, as well as federal prosecutors from the District of Columbia and the Justice Department's Counterterrorism Section—expended hundreds of thousands of investigator work hours on this case. Their efforts involved more than 10,000 witness interviews on six different continents, the execution of 80 searches, and the recovery of more than 6000 items of potential evidence during the course of the investigation. The case involved the issuance of more than 5750 grand jury subpoenas and the collection of 5730 environmental samples from 60 site locations.[13]

In 2008, days after USAMRIID scientist Bruce Ivins committed suicide, the FBI publicly named him as the sole perpetrator. The bureau formally closed the Amerithrax case in 2011 though skeptics, including members of Congress, have called for further inquiries on the matter.[14] Apart from this controversy, FBI scientists had played key roles in the investigation.

The bureau's Hazardous Material Response Unit was instrumental in the immediate response to the anthrax attacks. The unit had been formed by two FBI scientists, Randall Murch and Drew Richardson, during preparations for the 1996 Olympics.[15] Its mission is to respond to criminal incidents involving hazardous materials, including the use of biological agents, while maintaining protocols for safety and evidence gathering.[16]

In addition to the traditional investigative procedures, the case brought to bear the collaborative work of almost one hundred analytic scientists: geneticists, chemists, physical chemists, and microbiologists. They were drawn from various federal agencies, national laboratories, academic institutions, and industrial settings. The field of microbial forensics, a term coined in the late 1990s, was further formalized in 2003 in a seminal article by Budowle and colleagues:

> [S]cientists can play a substantial role in thwarting the use of bioweapons by developing tools to detect and to determine the source of the pathogen and to identify those who use such biological agents to create terror or to commit crime. By developing a robust microbial forensics field, security can be enhanced beyond physical locks and barriers.[17]

Infectious diseases are invisible and mysterious. There is no blast, no sound, no cloud or plume, no observable effect upon release of a biological agent. The nature of the organisms, such as bacteria or viruses, and the diseases they cause, is complex and requires training to understand. At the same time, there is a universal repugnance for the use of disease by one human being to kill or maim another.

Thus, when speaking at the Meeting of States Parties to the BWC in December 2009, Under Secretary for Arms Control Ellen Tauscher stated that "the Obama administration's new strategy for countering biological threats—both natural and man-made—rests upon the main principle of the BWC: that the use of biological weapons is repugnant to the conscience of mankind."[18]

THREE KEY BIORESPONDER GROUPS

In reviewing the timeline of the anthrax attacks,[19] three groups of players with key roles in the immediate and long-term response emerge (see Table 15.1). "The first responders will not be traditional ones such as police, fire, and emergency medical services, but instead will be physicians in their offices or in emergency rooms, and the public health system," said Margaret Hamburg, who was assistant secretary for planning and evaluation at the Department of Health and Human Services from 1997 until 2001.[20]

We focus here on three key science-responder groups and the systems they had at their disposal, followed by improvements and remaining challenges:

1. Laboratorians, who are part of the Laboratory Response Network (LRN)
2. Public health staffers (epidemiologists, pathologists, other medical personnel) in the US Public Health Service, who track the outbreak and assist in making decisions about treatment, quarantine, and related issues
3. Basic scientists, who strive to develop methods for (1) microbial forensic investigation and (2) outbreak detection and diagnosis

The Laboratory Response Network

The LRN was established in 1999 under the auspices of the CDC and the FBI and in accordance with Presidential Decision Directive 39.[21] Its mission is to "maintain an

TABLE 15.1   **Three Groups of Players with Key Roles in the Immediate and Long-Term Response**

Field	Profession	Tasks
Healthcare	MD Nurses First responders Laboratorians	Diagnosis Treatment
Public health	MD PhD Laboratorians EIS officers	Epidemiology Coordination of surveillance
Science	Microbiologists Chemists Physical chemists	Diagnosis Detection Countermeasures

integrated national and international network of laboratories that are fully equipped to respond quickly to acts of chemical or biological terrorism, emerging infectious diseases, and other public health threats and emergencies."[22] There are three structural layers of the LRN, the broadest of which comprises the "sentinel" laboratories, certified laboratories usually found associated with hospitals. These 25,000 private and commercial laboratories are engaged in diagnostic microbiology, using commercially available diagnostic tests. The sentinel laboratories have policies and procedures in place for referral of specimens.

The middle layer represents the "reference laboratories," about 160 of them, in the United States, Canada, Australia, and the United Kingdom. These include state and local public health, federal, military, veterinary, agricultural, environmental, and food-testing laboratories. Through validated tests on diagnostic samples, these facilities are equipped to confirm or rule out the presence of a bioterrorism agent. The specific organisms on the "select agent" list are periodically reviewed and can be found on the CDC website.[23]

At the apex of the system, there are the national laboratories, including those at the CDC and USAMRIID, that definitively characterize the samples provided by the reference laboratories. As the LRN was being rolled out, the CDC also developed a "critical agents" list, which includes organisms less pathogenic than conventional biological weapons agents, yet still capable of having a significant impact on public health, such as salmonella and shigella.

The LRN and the Anthrax Attacks   The LRN was barely 2 years in operation when Larry Bush, the astute Florida infectious disease physician, initiated the correct response to his suspicion about Robert Stevens' condition. Dr. Bush ("sentinel") contacted Dr. Jean Malecki, head of the County Health Department, who had the samples sent to the state health laboratory in Jackson ("reference"), which in turn contacted the CDC (national laboratory). The smooth and efficient process among these laboratories was enhanced by many previous interactions between Bush and Malecki; the entire diagnosis was carried out within 48 hours.

The introduction of the LRN was accompanied by increased education and training of sentinel laboratory personnel. Many laboratorians had never seen a case of anthrax or plague. The anthrax mailings were an educational experience for all and highlighted weaknesses as well as strengths in the system. Some of the problems, especially regarding communication failures, have been chronicled in Chapter 2 of this book as well as in other publications.[24] But by using LRN protocols already in place, state health departments worked cooperatively as part of the nationwide response.

Although only four states (New York, New Jersey, Connecticut, and Florida) and Washington DC were directly affected by spores in the mail, laboratories throughout the country were engaged in assessments of suspected specimens. Thus the experience was "more realistic than the scenarios that are generated for planned exercises."[25] In the ensuing decade, a series of exercises was developed jointly by the College of American Pathologists, the Association of Public Health Laboratories, and the CDC.[26] In these exercises, biothreat agents and non-biothreat organisms were provided to more than 1300 sentinel laboratories to test diagnostic capabilities. Accurate identification ranged from 83.8 to 99.9 percent among the laboratories, and with practice, the diagnostic time interval was reduced from as long as 10 days to only 3 days.

The LRN Going Forward   The LRN continues to be challenged by public health emergencies.[27] In addition to outbreak detection, the LRN is a key component of BioWatch, a surveillance program of the Department of Homeland Security that has positioned air sampling detectors in undisclosed locations in highly populated areas. The filters from these samplers are regularly removed and analyzed by designated laboratories.

The LRN is composed of allied health workers including clinical laboratorians and emergency physicians. Dr. Dan Hanfling, Special Advisor on Emergency Preparedness and Disaster Response to the Inova Health System, points out that while hospital emergency departments have long supplied a significant share of the public health safety net, their roles have now expanded to include that of community protector.[28]

The Public Health Service

The US public health infrastructure at the time of the anthrax attacks was not prepared for a disaster of this magnitude.[29] Reduced financial support for public health since the 1980s had resulted in severe deficiencies in both laboratory capacity and trained personnel. Some states did not even have a public health department. Many health departments lacked access to the Internet and e-mail; only 10% had any training in bioterrorism awareness, and just 20% of hospitals had bioterrorism response plans.[30]

The Clinton administration had tried to increase spending for bioterrorism preparedness and in 2000 it endorsed a bill to strengthen the federal role in public health preparedness.[31] But there was little congressional support for the effort and it remained underfunded at the close of the fiscal year.[32]

Public Health Response and the Anthrax Attacks   Despite a relatively efficient start with the diagnosis of the first case in Florida, the overall response to the outbreak was spotty. For example, when the FBI and the CDC descended on the postal distribution center in Hamilton, New Jersey, to begin their investigation, they neglected to inform the local police or town mayor in advance. Furthermore, there was little local public health capacity to assist in the decision-making process for treatment of the 1000 postal workers at risk of exposure. The area hospitals, as elsewhere in the United States, were "not integrated with the public health system but [operated] as independent, private sector enterprises."[33] It was Hamilton's mayor, Glen Gilmore, who independently organized and stocked an emergency clinic in a private hospital to distribute antibiotics. In the end, of the six cases of anthrax in New Jersey, two were inhalational and four cutaneous; but none died.

In all, some 30,000 people who might have been exposed to the attack spores were advised to begin a course of antibiotic prophylaxis. The antibiotic initially recommended was ciprofloxacin. The recommendation was later changed to doxycycline, which some saw as reflecting confusion about the proper response. Both antibiotics have side effects and many individuals discontinued treatment on their own. In addition to the antibiotics, in December 2001, the CDC offered the anthrax vaccine to members of high-risk groups—workers in the postal service, the senate office buildings, and the Florida media building where Stevens had been employed.

The vaccine was an investigational new drug and informed consent was required of anyone receiving an inoculation. Several postal workers declined the offer. Their reluctance was ascribed variously to lack of trust, risk perception, disagreement about the recommendation, and the controversy over the military's use of the vaccine. Some compared the vaccine trial to the mid-twentieth century Tuskegee syphilis study in which hundreds of black men, ostensibly receiving treatment for syphilis, were given placebos.[34]

Communication and Bioterrorism   As with all disaster planning, effective communication is key to a successful outcome, whether among responder groups or between officials and the public. Chapter 2 underscored the distinctive communication challenges generated by the 2001 anthrax attacks. But in any bio-attack, identification of the specific causative agent is essential. Wray and associates considered, for example, the differences between a non-communicable (anthrax) and communicable (plague) disease: "It is likely that the communicable nature of plague and the high rate of mortality due to the infection would cause even higher levels of panic. Effective communication in the form of consistent, accurate, and timely information disseminated efficiently across the media and government agencies may well be the key to preventing panic and potentially saving lives."[35]

As one analyst observed, in the event of a biological attack, "[communication] challenges will be greater than before, and press officers sent to site investigations will need clear guidelines for coordinating messages with those emanating from Atlanta or Washington so that alarmed citizens are not bombarded with conflicting advice."[36] Government at all levels needs rapid access to information to support critical decisions in public health. "The major barriers to information access include time, resource reliability, trustworthiness/credibility of information, and information overload."[37] In a bioterror event, rapid communication may depend on transmission via public health digital knowledge management systems. An acknowledged challenge to this dependency is the vulnerability of cyber-based systems to possible attack or failure, as discussed in Chapter 18.

Role of Scientists in the Investigation of the Anthrax Attacks

Once the cause of Robert Stevens' death was shown to be anthrax, the first task of scientists was to determine which of the hundreds of known strains of *B. anthracis* was the causative agent. The strain recovered from Stevens was immediately airlifted to Paul Keim, an expert on anthrax genomics at the University of Northern Arizona. Keim and his laboratory group had previously characterized a set of specific DNA markers located throughout the chromosome of *B. anthracis*. The pattern of markers was different for each of the strains they had studied. The attack sample was quickly determined to be the Ames strain, which had been used by government laboratories to test anthrax vaccines. A day after receiving the sample, on October 5, Keim notified the FBI and CDC of his findings.

The next challenge for the FBI was to determine which research laboratories possessed this strain. In 2001, there were few regulations governing possession of *B. anthracis* and other select agents. A research laboratory was required to register with the CDC only when transferring a sample to another laboratory; those already in possession of anthrax were not known or registered with any agency. The FBI began to interview researchers throughout the country and overseas.

In January 2002, the FBI sent a letter to the 42,000 members of the American Society for Microbiology (ASM). The letter described the presumed characteristics of the perpetrator and asked for help from the ASM members toward identifying suspects. Meanwhile, scientists in more than 25 national, academic, and commercial laboratories contributed to the analyses of the attack material. While physical and chemical studies did not appear to contribute significantly to attribution, the FBI relied heavily on genetic analyses of the attack material in its final statements regarding attribution.[38]

In 2011, the National Research Council of the National Academy of Sciences released a report describing and critiquing the scientific approaches used in the anthrax investigation. The report found that the genetic analyses were consistent with, but not unequivocal proof of, the FBI's conclusion that the attack strain was derived from a specific flask of spores maintained at the USAMRIID laboratory.[39] With the death in 2008 of Bruce Ivins, the FBI's prime suspect, the case may never be solved.

Role of Scientists in Detecting a Biological Threat: BioWatch

The federal response to the anthrax attacks included increased investment in countermeasures research and public health infrastructure. It also implicitly broadened the definition of responders to include laboratorians and scientists. Basic scientists in national laboratories, such as the Livermore and Sandia National Laboratories, have long fulfilled research and development functions. But the 2001 bio-attack prompted mobilization of these investigators to veer in new directions: work on the anthrax investigation and on the development of BioWatch, the early warning system of pathogen release.

The structure of BioWatch is complex. While launching, funding, and overseeing detector locations are carried out by the Department of Homeland Security, the program's components are directed by three different agencies. The Environmental Protection Agency monitors the sampling aspect; the CDC controls testing and analysis; and the FBI directs the response. The exact locations of the monitoring sites remain secret, but they are known to be present in the 30 largest US cities.

Scientists have played key roles regarding the BioWatch program: helping to set it up, maintaining it, and now critiquing it. Among the criticisms is that the program is labor-intensive and expensive. Several public health officials have expressed skepticism about the value of BioWatch in comparison with more traditional methods of epidemiological monitoring of disease outbreaks. The National Research Council released a report in 2010 that reviewed the successes and failures of the program and found deficiencies in management, priorities, and technical issues that affected the system's overall performance.[40] But whatever the scientists' criticisms, they will doubtless play a role in remediating these deficiencies as well.

BIODEFENSE AND PREPAREDNESS: LOOKING FORWARD

The effects of the anthrax attacks continue to play out on a national and international scale in medical research, public health, intelligence gathering, criminology, and medical ethics. The attacks identified weaknesses in the US public health system, but they also spawned many improvements. Health security became a congressional focus as well as a priority for the Department of Health and Human Services, which in 2010 released its National Health Security Strategy. Initiatives have led to the expansion of existing programs, such as the Laboratory Response Network and the Strategic National Stockpile, and the creation of new alliances, as with the incorporation of public health components into the National Response Framework and the National Incident Management System.[41]

For much of the past decade, increased funding and infrastructure support at the state level enhanced response time and communication during public health emergencies. However, in recent years, economic pressures, coupled with complacency, have posed a danger to continuing national health preparedness. Between 2008 and 2010, more than 44,000 jobs were lost in state public health sectors.[42] The fields of public health, clinical care, and laboratory science are all essential to biosecurity. Weakening any of these areas can only weaken the nation's biodefense.

NOTES

1. Cole LA. *The Anthrax Letters: A Bioterrorism Expert Investigates the Attacks that Shocked America*. Revised. New York: Skyhorse Horse Publishing; 2009. Guillemin J. *American Anthrax: Fear, Crime, and the Investigation of the Nation's Deadliest Bioterror Attack*. New York: Henry Holt; 2011.

2. Cole LA. 2009.

3. MMWR 1976;25:33, 34.

4. Cole LA. 2009.

5. U.S. House of Representatives, Subcommittee on Homeland Security. Hearing on Emergency Preparedness and Response. May 13, 2011.

6. Martin M. Obama administration gets an "F" for bioterrorism defense. Global Security Newswire, January 26, 2010.

7. National Academy of Sciences, National Research Council. *Review of the Scientific Approaches Used during the FBI's Investigation of the 2001 Anthrax Letters*. Washington, D.C: National Academies Press; 2011.

8. The experiments involved the generation of host range mutants of avian-specific influenza H5N1 virus, known to be associated with a high-fatality rate (50–60%) for humans. The new mutants were capable of infecting ferrets, a standard model for human host, and of transmission by the aerosol route. The novel viruses are potentially extremely dangerous for the human population. At issue is whether the details of the mutations leading to the novel phenotypes should be published. As of January 2012, a 60-day moratorium was declared on continuation of the research. Fouchier RAM, Garcia-Sastre A, Kawaoka Y, and 36 coauthors. Pause on avian flu transmission studies. Nature Online, January 20, 2012. Available at http://www.nature.com/nature/journal/vaop/ncurrent/full/481443a.html#auth-3, accessed June 21, 2012.

9. Sepkowitz KA. Anthrax and anthrax anxiety: Sverdlovsk revisited. Int J Infect Dis 2001;5:178–179. Meselson M, Guillemin J, Hugh-Jones M, et al. The Sverdlovsk anthrax outbreak of 1979. Science 1994;266:1202–1208.

10. Presidential Decision Directive 39: U.S. Policy on Counterterrorism, June 21, 1995.

11. The Public Health Threats and Emergencies Act of 2000 (P.L. 106-505) was the first bill passed into law that addressed domestic bioterrorism-related public health needs.

12. Blueprint for the National Domestic Preparedness Program. Available at www.securitymanagement.com/library/ndpo1201.pdf, accessed January 21, 2012.

13. Amerithrax or Anthrax Investigation. Available at http://www.fbi.gov/about-us/history/famous-cases/anthrax-amerithrax/amerithrax-investigation, accessed January 21, 2012.

14. Congress to examine FBI anthrax investigation at Holt's request. Available at http://holt.house.gov/index.php?option=com_content&task=view&id=600&Itemid=18, accessed January 21, 2012.

15. http://www.fbi.gov/about-us/lab/hmru, accessed January 21, 2012.

16. Ryan J, Glarum J. *Containing and Preventing Biological Threats*. Burlington, MA: Butterworth-Heineman; 2008.

17. Budowle B, Schutzer SE, Einseln A, et al. Public health. Building microbial forensics as a response to bioterrorism. Science 2003;301(5641):1852–1853.

18. Tauscher E. Preventing biological weapons proliferation and bioterrorism. U.S. Department of State. Address to the Annual Meeting of the States Parties to the Biological Weapons Convention, Geneva, Switzerland, December 9, 2009. Available at http://www.state.gov/t/us/133335.htm, accessed February 10, 2012.

19. For timelines of both the factual events and the scientific developments, see National Academy of Sciences, National Research Council. Review of the Scientific Approaches.

20. Thomas P. *The Anthrax Attacks: A Century Foundation Report*. New York: Century Foundation; 2003.

21. Presidential Decision Directive 39.

22. Centers for Disease Control and Prevention. The Laboratory Response Network Partners in Preparedness. Available at http://www.bt.cdc.gov/lrn/, accessed September 30, 2011.

23. Centers for Disease Control and Prevention. National Select Agent Registry. Available at http://www.selectagents.gov/select%20agents%20and%20Toxins%20list.html, accessed January 21, 2012.

24. Kahn LH. *Who's in Charge? Leadership during Epidemics, Bioterror Attacks, and Other Public Health Crises*. Santa Barbara, CA: Praeger Security International; 2009.

25. Nolan PA, Vanner C, Bandy U, et al. Public health response to bioterrorism with *Bacilllus anthracis*: Coordinating public health laboratory, communication, and law enforcement. J Public Health Manag Pract 2003;9:352–356.

26. Wagar EA, Mitchell MJ, Carroll KC, et al. A review of sentinel laboratory performance: Identification and notification of bioterrorism agents. Arch Pathol Lab Med 2010;134:1490–1503.

27. Centers for Disease Control and Prevention. Examples of the Laboratory Response Network (LRN) in Action. Available at http://www.bt.cdc.gov/lrn/examples.asp, accessed January 21, 2012.

28. Hanfling D. Public health response to terrorism and bioterrorism: Inventing the wheel. Trust for America's Health, August 4, 2011. Available at http://healthyamericans.org/health-issues/story/public-health-response-to-terrorism-and-bioterrorism-inventing-the-wheel-by-dan-hanfling, accessed February 10, 2012.

29. Frist B. Public health and national security: The critical role of increased federal support. Health Affairs 2002;21:117–130.

30. National Association of County and City Health Officials [NACCHO]. Information Technology Capacity and Local Public Health Agencies. Research Brief. Washington, DC: NACCHO; 1999.

31. The Public Health Threats and Emergencies Act of 2000 (P.L. 106-505).

32. Frist.

33. Kahn LH, Barondess JA. Preparing for disaster: Response matrices in the USA and UK. J Urban Health 2008;85:910–922.

34. Quinn SC, Thomas T, Kumar S. The anthrax vaccine and research: Reactions from postal workers and public health professionals. Biosecur Bioterror 2008;6:321–333.

35. Wray R, Jupka K. What does the public want to know in the event of a terrorist attack using plague. Biosecur Bioterror 2004;2:208–215.

36. Thomas.

37. Revere D, Turner AM, Madhavan A, et al. Understanding the information needs of public health practitioners: A literature review to inform design of an interactive digital knowledge management system. J Biomed Inform 2007;40:410–421.

38. Federal Bureau of Investigation. Amerithrax or Anthrax Investigation. Available at http://www.fbi.gov/about-us/history/famous-cases/anthrax-amerithrax/amerithrax-investigation, accessed January 21, 2012.

39. National Academy of Sciences, National Research Council. Review of the Scientific Approaches.

40. Institute of Medicine and National Research Council Report. *Effectiveness of National Biosurveillance Systems: BioWatch and the Public Health System*. Washington, DC: Institute of Medicine; 2010.

41. Khan AS. Public health preparedness and response in the USA since 9/11: A national health security imperative. Lancet 2011;373:953–956.

42. Ibid.

16

THE ROLE OF THE LAW ENFORCEMENT OFFICER

Graeme R. Newman and Ronald V. Clarke

The terrorist attack on the US World Trade Center Twin Towers on September 11, 2001, is commonly compared to the Japanese attack on Pearl Harbor in 1941 in which 2401 persons were killed, more than 1000 wounded, and the US Pacific Fleet decimated. The shock to the nation was such that it pulled the isolationist-leaning United States into World War II. The shock of the 9/11 attack, which killed more than 2500 persons and continues to cause deaths, injury, and sickness in its aftermath, pushed the United States into wars in Afghanistan and Iraq. The truly common element of the two disasters was the surprise of the attacks and the chaos that followed. In hindsight, critics have argued that the US military commanders should have anticipated and been prepared for the Japanese attack. The 9/11 Commission has similarly criticized the US counterterrorism agencies for their failure to anticipate and therefore prepare for attacks such as that of 9/11.

9/11, THE BEGINNING

In what follows, we provide a brief account of the events that occurred on the ground during the 9/11 attack on the World Trade Center (WTC). The account is based on the exhaustive report provided by the 9/11 Commission.[1] Our review is confined to just the first 17 minutes of the attack. It may be argued that the 1993 attack on the WTC should also be considered part of a case study of this disaster. But several terrorist attacks had been directed against US targets in the intervening years, such as those against US embassies abroad and against the USS Cole. Their consideration would mean broadening our focus beyond that of local police to include the counterterrorism policy, strategy, and operations of federal agencies such as the FBI, CIA, and others. Be that as it may, this case study of local police and other first responders to the one 9/11 attack occurs against a backdrop of counterterrorism policies

Local Planning for Terror and Disaster: From Bioterrorism to Earthquakes, First Edition.
Edited by Leonard A. Cole and Nancy D. Connell.
© 2012 John Wiley & Sons, Inc. Published 2012 by John Wiley & Sons, Inc.

and practices of many federal agencies, which were, for the most part, the agencies most criticized by the 9/11 Commission. Local first responders were spared the tough criticism leveled at federal agencies because their actions were overshadowed by the heroism of many first responders, bystanders, and victims who lost and risked their lives.

Setting the Scene

The unfolding of the 9/11 attack is extremely complex because there were multiple events occurring in different places, so each event was unfolding according to its own timetable. Along with the thousands of 911 calls made by victims, there were competing first responder organizations transmitting information and issuing orders: the North American Aerospace Defense Command (NORAD), Federal Aviation Administration (FAA) and its various subsidiaries, the federal counterterrorism task force, Secret Service, various military departments, local city police and fire departments, medical emergency personnel, and various executives such as the office of the Vice President. Finally, there appeared to be no mechanism to employ any existing disaster team or plan to manage this complex attack. This, even though several first responder organizations had recently undergone training routines for responding to a major terrorist attack, one even using the scenario of a commercial plane crashing into a building.[2]

The First 17 Minutes

8:46 A.M.   Flight 11 crashes into North Tower. Emergency 911 number is swamped with calls; 911 operators advise occupants of North Tower to stay put and wait for rescue workers, according to standard operating procedure. Most occupants begin evacuating regardless of instructions.

8:49 A.M.   South Tower deputy fire director informs occupants that the building is safe and that they should remain in their offices.

8:50 A.M.   New York Police Department (NYPD) dispatches helicopters to survey damage. No Fire Department of New York (FDNY) officers are placed in helicopters as per standard operational procedures. NYPD determines rooftop rescue not possible. Information not conveyed to FDNY.

8:52 A.M.   FDNY arrives at site. Dispatchers have no information about location or magnitude of the impact zone. FDNY chiefs in the tower lobby determine that all occupants should evacuate. This information is not conveyed to 911 operators or FDNY dispatchers. Units ordered to climb tower to impact zone. FDNY advised that building not likely to collapse. Chiefs speak with Port Authority Police Department (PAPD) and Mayor's Office of Emergency Management.

8:57 A.M.   FDNY issues evacuation order for South Tower and is unable to keep track of various rescue operations and personnel deployment. NYPD clears major thoroughfares, supervises evacuation of people at WTC plaza. FDNY works with PAPD to evacuate WTC plaza.

9:00 A.M.   PAPD advises FDNY that full evacuation orders have been issued through the public address system, but this is not fully functioning. PAPD officers on the scene help in rescue operations on lower floors. Not all officers have WTC command radios. PAPD commanding officer orders evacuation of all civilians from WTC complex.

9:03 A.M.   Flight 175 hits South Tower. 911 operators are again overwhelmed with calls; they advise occupants to stay put. FDNY analog radios function poorly because WTC repeater system not switched on properly. NYPD sends small rescue teams up towers. PAPD officers respond individually; PAPD does not know how many officers are responding or where they were going.

Understanding What Happened

Overall, it took these 17 minutes for first responders to understand that this was a rescue mission, not a firefighting mission. Thirty-nine minutes later the South Tower collapsed and in a further 29 minutes the North Tower collapsed.

At the local level, the inability to communicate clearly and efficiently severely affected actions on the ground. It was extremely difficult for responders to comprehend the enormity of the 9/11 calamity. Assessing the severity of an attack can be the greatest challenge to a first responder team. Without some knowledge of how serious the disaster is, it is difficult to deploy personnel and equipment effectively and difficult to assess the danger involved for the first responders themselves. The response time of the NYPD and FDNY to the attack was extremely rapid (some PAPD personnel were on the scene when the incident occurred). But it was of little use getting there so quickly if there was no means of properly assessing the nature and extent of the damage, and the numbers of rescue personnel required. Without such information a triage protocol could not be put in place.

Both the NYPD and FDNY did in fact have different levels of alarm call-up, but these proved too general and, particularly in the case of the FDNY, resulted in the congregation of many personnel at the scene, without a sense of how to deploy efficiently. Many simply walked up the towers as the occupants were walking down. The result was that many first responders, having achieved little, became exhausted from climbing and remained trapped in the towers when they collapsed.

The Mayor's Office of Emergency Management played a limited role in directing the operations, even though coordinating the first responder agencies was an explicit part of its mission. Initially, it did make calls to the Federal Emergency Management Agency (FEMA) requesting rescue teams and to the Greater Hospital Association. Fatefully, the Office of Emergency Management headquarters was on the 23rd floor of 7 World Trade Center.

The Information Problem

The most widely criticized aspects of the response to the 9/11 disaster were the inadequate and even nonfunctioning communications equipment and the lack of coordination among the various first responder organizations. Both these factors contributed to inaccurate and conflicting information passed among first responders. Figure 16.1 is a highly simplified representation of information flow and communications during the first 17 minutes. There was a multiplicity of information sources. We can see from the chart that the direct links of the 911 call center to victims and agency dispatchers were crucial. As a source of information they were possibly the most important, since they could get information from

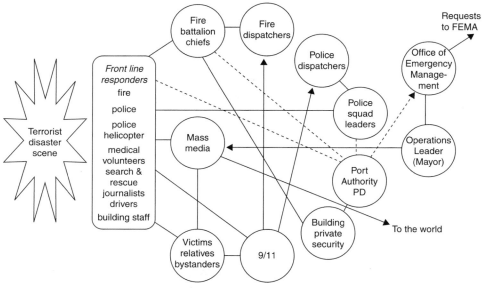

- All lines are bidirectional unless otherwise indicated
- Broken line indicates sporadic information flow
  because of poor communications technology.

FIGURE 16.1   Simplified communications diagram, 9/11 attack, first 17 minutes. (Adapted from Newman GR, Clarke RV. *Policing Terrorism: An Executive's Guide*. Washington, DC: USDOJ Center for Problem-Oriented Policing; 2008. Brief 45.)

victims at the scene, plus convey information to those same victims, or subsequent callers. They could also convey this information to the command center of the emergency: in this case the Office of Emergency Management, which, if it were functioning properly would then collate the information from other sources and make sure accurate information was then transmitted down the command chain (i.e., to NYPD, FDNY, etc.) and back to the 911 call center. The flow chart also reveals the links that are missing between agencies. These links failed because of barriers (organizational and technological) that made it difficult for information flow to move across them.

Advanced communication technologies such as cell phones, enhanced 911, various radio band technologies, and mass media radio and TV have enormously increased the ability of people to communicate. The problem is that these technologies do not care whether the information being transferred is true or accurate. This is why disaster management control centers, such as the Los Angeles County Control Center that copes with major forest fires every year in California, maintain a close link with the media to make sure that reportage is accurate, does not exaggerate the disaster, and does not add to the problems faced by first responders. Sometimes the controls break down and the results are not always as feared. For example, the major network news reports during Hurricane Katrina seem to have portrayed a more accurate and complete picture of conditions in New Orleans than the information passing through official channels between FEMA officials and local government officials.

In the case of the WTC attack on 9/11 the media played an additional role: it was the conduit for the Mayor to convey reassurance and encouragement to the community. As Figure 16.1 illustrates, the Mayor's place in the flow of information was essentially linked to the mass media, and secondarily to the Office of Emergency Management. As it happened, the role of that office, which had been established many months before 9/11, was to overcome all impediments to communications among the different departments of first responders and to assume the role, during a disaster, of coordinating the entire operation. Unfortunately, as the 9/11 Commission report clearly shows, the NYPD and FDNY continued to operate as independent agencies. Their radios could not talk to each other, nor did their users want to talk to each other.

The Organizational Problem

There was poor coordination among the various first responder organizations and there was no single commander of the overall operation. Voluminous accounts have been offered about chain of command in facing disasters, many of which are encapsulated in the National Incident Management System (NIMS, see Fig. 16.2). This model is widely used throughout

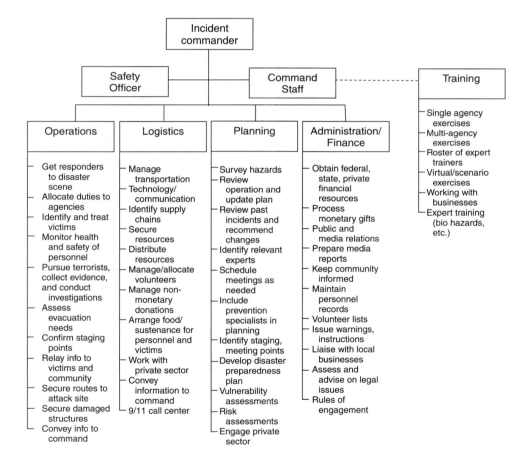

FIGURE 16.2   National Incident Management System (NIMS). *Note:* This is a simplified version adapted to match the needs of local police. There are many versions and updates of the system, the more recent being HPD-5 disseminated by the Department of Homeland Security. (Adapted from Newman GR, Clarke RV. *Policing Terrorism: An Executive's Guide*. Washington, DC: USDOJ Center for Problem-Oriented Policing; 2008. Brief 45.)

the Unites States to impose a uniform and cooperative approach on first responders. It originated from problems in dealing with large fires in the 1970s, which demanded responses from more than one agency and more than one jurisdiction. There were several factors that made fighting these fires difficult: unclear command structures, misunderstandings among responder agencies, no means of coordinating each agency's tasks, lack of a clear leader, and competition for scarce resources among responder agencies.

The sensible way to overcome such difficulties is to establish a clearly structured organizational plan that ensures that everyone knows who does what. This is what NIMS does, and it was operational in New York City well before 9/11 through the Mayor's Office of Emergency Management. However, the Mayor's emergency management staff remained in the background during the 9/11 disaster. As far as one can tell from the 9/11 report, Mayor Giuliani did not take control of the operations. He did, however, appear as the leader of the city's response to the attack on that day and the days that followed. In fact, it was generally concluded that his public speeches and actions played a major role in helping people comprehend the magnitude of the disaster, and cope with the loss and disturbances to their everyday lives resulting from the attack.

It might be argued that the Mayor's emergency management staff should have played a stronger role in coordinating the first responder agencies, but this would have been hampered by the incompatible communications technologies used by the different first responder agencies. A single operational commander might have been able to mitigate some of the competition and lack of coordination between the fire and police departments, but given the circumstances on the ground, especially in the first 17 minutes, it is unlikely. When an attack is in progress, quick decisions have to be made, often based on flimsy information. There must, therefore, be a way to move beyond the static NIMS organizational chart in order to respond to how things change on the ground.

The crucial part of any command structure in action is the collection and movement of information from the ground to the commander and down again. The NIMS chart shows nothing of this dynamic need for operational efficiency. Nor does it indicate from what participating organizations the various commanding officers come. Should the incident commander be the police chief, the fire chief, or the emergency medical director? The quick answer to this question is that it would depend on the type of disaster. A protocol for assessing what type of disaster would require what kinds of commanders and in what capacities is part of the overall disaster management plan that most local communities now have in place. We assume for the purposes of this narrative that the prime responsibility for responding to and preparing for a terrorist attack lies with local police, though many other organizations are and should be involved at various stages of response.

The Aftermath

The aftermath of the attack unfolded in three stages roughly as do most disasters. In the first immediate stage Mayor Giuliani, President Bush, and other public officials helped tremendously to buoy up the spirit of the people of New York and the nation as a whole. There was an outpouring of sentiment and massive influx of volunteers to help. The NYPD devoted many months identifying remains of victims, contacting families, compiling lists of the missing, killed, and injured.

The second stage began fairly quickly as questions were asked: Why were we not prepared for such an attack? Whose responsibility was it to prepare for and anticipate such threats? How could these terrorists carry out such a complex operation right under the noses of law enforcement? The 9/11 Commission compiled a detailed account of the circumstances that led up to the attack. It was especially critical of the law enforcement organizations at the national level that failed to uncover the plot, and faulted the CIA and FBI in particular for failing to share information concerning suspected terrorist activity. Although the Commission made clear that there were serious defects in the emergency management preparedness of local police and fire departments, it restrained its criticisms because of the outpouring of grief and sentiment for those first responders who lost their lives and who were the heroes of the day. Actually, from a formal point of view the local agencies seemed prepared: the Mayor had set up an emergency management team more than a year before the attack and there had been disaster response training sessions, including one that used the scenario of an airplane attack on the WTC. Instructions had even come from the Mayor's office for departments to update communications equipment that allowed for interoperability.

The third stage, which still continues, has been mired in controversy and recrimination. Congress, among other things

1. Created the new Department of Homeland Security, placing a variety of law enforcement agencies (e.g., FBI, immigration, customs) and disaster response components (e.g., FEMA, Office of Emergency Communications) under its umbrella. The justification was that this reorganization would enhance information sharing among departments.

2. Allocated millions of dollars to communities for protection against terrorist attacks. Unfortunately, this money was often allocated according to political largess rather than assessments of those communities most at risk of attack.

3. Established procedures (still controversial) for the legal processing and internment of terrorist suspects.

4. Provided extra powers (still controversial) to federal agencies to intercept communications and maintain surveillance of suspected terrorists or their accomplices.

The 9/11 Commission released a report card of achievements to coincide with the 10th anniversary of the attack. It noted that state and local government organizational and communications interoperability still needed improvement, and that federal unity of command was still lacking.[3]

Of particular interest to criminologists has been the ongoing debate fostered by the Obama administration regarding legal jurisdiction for alleged terrorists. The administration has insisted on treating terrorist acts as criminal acts, which therefore requires the standard procedures of criminal justice, including due process, trial in federal court, and defense attorneys. From the point of view of local police, we believe that thinking of terrorism as crime is a useful way to formulate the problem from an operational and planning point of view, as we will elaborate shortly. (The legal controversies are perhaps relevant to federal authorities, but less so for local police.)

From the point of view of local police, the most significant event that followed the 9/11 Commission report was that the NYPD set up its own counterterrorism force in an attempt to collect its own information—in effect to do what the FBI and CIA had failed to do.[4] Events over the past 10 years suggest that this may have been a sensible move, given that the NYPD has thwarted a number of serious terrorist attacks. Commensurate with this move was the criticism of the federal government by former NYC police chief William

Bratton, that it had failed to recognize the significant and crucial role of local police in preparing for and responding to terrorist attacks.[5] The 9/11 Commission's solution was to strongly recommend the sharing of information across agencies, which resulted in a radical reorganization of federal agencies and creation of a completely new department of Homeland Security. But, while maintaining a public face of close cooperation with the FBI, in fact tensions have sometimes surfaced between NYPD and the FBI, most recently in October 2011 with the arrest by NYPD of a terrorist suspect, Jose Pimental, whom the FBI publically pronounced as not a threat.

UNDERSTANDING TERRORISM FROM THE LOCAL POLICE PERSPECTIVE

The fields of Problem-Oriented Policing[6] and Situational Crime Prevention[7] have, over the last two decades, developed an effective way to assess crime and disorder problems to construct responses for reducing or preventing them from recurring. The prime focus is to identify the opportunities of which criminals take advantage. Reduce or eliminate these opportunities, and the problem is potentially solved. This approach can be applied just as easily to terrorism if one thinks of terrorism as simply another form of crime.

Thinking of terrorism as a form of crime ("crime with a political motive") strips the concept of its mystique and international intrigue and counters the idea that it is a problem in the realm of spies and international politics. It is true that officially designating terrorism as crime brings with it some political, legal, and extra legal problems. But from the point of view of prevention (anticipating terrorist attacks, and especially their likely targets) special classification seems unhelpful. Calling terrorists foreign assailants, abiding by the Geneva Convention, and so on, has little relevance to preventing such attacks at the local level.[8] The relevance of such labels is largely confined to the processing of terrorists once caught. Of prime importance in anticipating and preventing terrorist attacks is to be ready if attacked, and even more important as part of that readiness, assess the likelihood of attacks, especially the targets that may be chosen. Experienced police will recognize that this is, or should be, part of the everyday operations of police who see it as their prime mission to prevent crime in their neighborhoods.

Some argue that terrorists are different from criminals because they are fanatics or ideological extremists, so local police must adopt a different approach to such determined offenders. This is wrong. No matter what their motivation, whether criminals or terrorists, all must be committed to their particular operation. They must make many choices in order to carry out their attacks or crimes: they must choose victims or targets, they must choose where they will approach their targets, they must choose how they will get to the place of operations, they must choose (in most cases) an escape route. It follows that understanding how terrorists carry out their tasks will provide police ways to intervene in their operations and make it harder for them to accomplish their missions. It matters little whether we call what they do terrorism or crime. The overall goal of local policing, therefore, is to reduce opportunities for crime, and thus terrorism.

Countering Terrorism: The Four Pillars of Terrorist Opportunity

Professors Clarke and Newman have identified four pillars of terrorist opportunity: targets, weapons, tools, and facilitating conditions.[9] These are often opportunities that are present in local communities (though this may vary according to type of attack). It follows that local police, if trained to identify and assess the availability of these opportunities, are key to protecting local communities from attack. And as we shall see further, this process of identifying opportunities is also key to mitigating the effects of an attack should one occur.

Targets   Targets attract terrorists differently. Their selection is probably the most important decision a terrorist will make. Their attractive characteristics may be summarized by the acronym EVIL DONE:

Exposed	The Twin Towers were sitting ducks to air attack.
Vital	The Twin Towers were the hub of US commerce.
Iconic	The Twin Towers were the icon of American capitalism.
Legitimate	Terrorists' sympathizers cheered when the Twin Towers collapsed.
Destructible	The Twin Towers were certainly destructible to the right weapon.
Occupied	The Twin Towers were full of office workers and first responders.
Near	Al Qaeda established its base just across the river in New Jersey.
Easy	Once the right weapon was found, destruction of the Towers was easy compared to the failed attempt with a truck bomb in 1993.

Weapons   Different weapons achieve different goals and are more appropriate for particular targets, and their preference to terrorists becomes more apparent in repeated attacks. The Twin Towers attack was a one-off attack, using a one-off weapon. Weapons characteristics that will affect terrorist choice are summarized by the acronym MURDEROUS:

Multipurpose	Commercial airplanes get you directly to the target, and they are highly incendiary.
Undetectable	Who would have thought that a commercial airplane was also a weapon?
Removable	The weapons of terrorism must be portable, usually relatively light and small. Commercial airliners are the exception, though they proved very portable by a sufficiently trained operative.
Destructive	Explosive devices have a greater kill rate than guns targeted at specific individuals. The Towers were a very big target, so a very large weapon was needed.
Enjoyable	Terrorists are clearly attracted to their weapons, especially those that make spectacular explosions.
Reliable	If terrorists have used a weapon, or one like it, many times before, they are likely to favor that weapon over another. Subsequent attempts to repeat use of an airliner have failed because of lack of practice and familiarity (e.g., the underwear bomber, shoe bomber).
Obtainable	Prior to 9/11 it was relatively easy to learn to fly an airliner in the United States.
Uncomplicated	Flying an airliner demands considerable skill, so it will be reserved for a one-off attack.
Safe	For repeated attacks, weapons need to be safe and predictable, unless of course, the terrorist plans also to kill himself.

Tools   All terrorist attacks depend on the tools of everyday life. Without such tools it is much harder for terrorists to reach a target or use their weapons. For many of the commonest attacks, such as car or truck bombings, drive-by shootings and targeted assassinations, terrorists are likely to need most of the following:

- cell phones or other means of communication;
- cars or trucks to transport themselves and weapons;

- cash or (false) credit cards, bank accounts or other means of transferring money;
- documents (false or stolen)—for example, drivers' licenses, passports or visas, and vehicle registration documents;
- maps (and increasingly GPS), building plans and addresses, so that the target location can be pinpointed;
- video and still cameras for surveillance;
- Internet access to collect information on street closures, traffic patterns, weather conditions, local news, and disseminate propaganda worldwide.

It may seem obvious that terrorists would need these everyday tools, but it is not obvious how they can use them without divulging at some point who they are and what they are up to. These tools are certainly widely available, but their visibility is also considerable. Using cash instead of a credit card to rent a car, for example, draws unwanted attention to the terrorist. Using a credit card exposes one's identity—unless it is stolen. For these reasons terrorists steal many of the everyday tools that are widely available, which of course opens them up to risk of getting caught. This is why there is a considerable, and increasing, overlap between terrorism and "traditional crime" such as money laundering, drug trafficking, and even bank robbery. It is obvious that a well-developed police–business partnership to maintain vigilance over these activities will help prevent not only terrorist attacks but ordinary crime as well.

Facilitating Conditions   Targets, weapons, and tools exist within physical, economic, and social environments. These environments serve to enhance their use; otherwise, they would be useless. At particular points in time or in particular regions or places, conditions may arise that facilitate terrorist ability to exploit these opportunities. These conditions make it ESEER from the terrorists' point of view because they make it:

Easy	When local officials are susceptible to corruption
Safe	When ID requirements for monetary or retail transactions are inadequate
Excusable	When family members have been killed by local antiterrorist action
Enticing	When local culture/religion endorses heroic acts of violence
Rewarding	When financial support for new immigrants is available from local or foreign charities

PREVENTING AND MITIGATING TERRORIST ATTACKS: THE ROLE OF THE LOCAL POLICE

Planning and Partnering

Identifying the opportunities for terrorists in the local community is the first step toward prevention and an enormous step toward mitigation. Only some of the above brief overview of opportunities will be relevant to a particular community. However, it is obvious that local police in many respects do not have the expertise or resources to conduct this kind of exercise on their own. Because we know that businesses are the most common (direct or indirect) victims of terrorist attacks, many occurring in places owned and operated by the private sector (the Twin Towers for example), it follows that police should develop working partnerships with business to help assess the vulnerability of targets in their communities. As we have seen in the previous section, not all targets are equally attractive to terrorists, so not all targets in a community need to be protected to the same degree. In sum, local police must partner with businesses and other community organizations in order to

- conduct risk assessment of local targets;
- assess vulnerability of targets;
- assess the likely injury and damage if a target is attacked;
- ensure that there are no conditions facilitating terrorist activity;
- monitor the trade in common tools of terrorism;
- develop a trusted system of information sharing among police, businesses, and other community organizations.

All of these activities are counterterrorism operations, but if carried out they also make for very effective everyday policing. They bring with them dual benefits: preventing terrorism and preventing crime.

Mitigating

We did not include in the previous section on planning and partnering the usual disaster response training which has become the most widely adopted mode of preparing for terrorist attacks and other disasters. While it is important for the various first responders to a disaster, such as police, fire, and medical emergency personnel, to have practiced together, if the preventive partnerships are well established as described earlier, this takes us more than halfway there. Most disaster training is conducted through workshops that are unavoidably artificial. They use scenarios, practice runs, and so on, but they are infrequently conducted and are expensive in terms of time and resources. And because they are infrequent they do not provide the ongoing daily familiarity that develops when police and community partners work together over a period of years. As noted earlier, the New York Mayor's Office of Emergency Management conducted scenario training for first responders prior to 9/11. But it had little impact on how each of these agencies responded to the attack.

Reducing Terrorist Rewards

A major part of terrorist reward is its aftermath. This is why Al Qaeda and other terrorist groups do their best to conduct simultaneous attacks making it difficult for first responders. Moreover, they often plan secondary attacks against first responders after they arrive on the scene. So we look upon the work to reduce the impact of a terrorist attack as itself a part of prevention. The severity of impact can be reduced if prevention responses have already been put in place. If terrorists cannot reach their target of choice, their planning and operations are put in jeopardy, and the impact of their attacks will be similarly reduced.

CONCLUSION

Of the four pillars of opportunity, target protection has a direct impact on disaster mitigation because

- it reduces terrorist reward
- ensures well-oiled communications and interorganizational operations because of preplanning responses focused on prevention

- businesses, schools, hospitals, community organizations already work together because of prevention responses and planning
- ensures long-term community support for disaster recovery

These real activities (not artificial training exercises) serve as the focal point for local police to accomplish specific objectives in terrorism prevention and disaster planning. It also lays the groundwork for identifying and reducing elements of the remaining three pillars of opportunity: weapons, tools, and facilitating conditions. In working with local businesses and community organizations, police can emphasize to their partners important actions they can take in continuing vigilance for possible terrorist activity and assist local police to develop in the collection of relevant intelligence, for example,

- Working with immigrant communities that inadvertently provide cover for terrorist activities, such as money transfers, opening of bank accounts
- Working with immigrant communities to identify individuals expressing or acquiring extremist ideologies, or are becoming radicalized
- Working with local banks to identify money laundering activities
- Working with retailers to track purchases of bomb-making materials
- Working with banks and retailers to monitor the incidence of credit card fraud

While "undercover" officers may be able to obtain this kind of information, much is also available from local citizens who come across relevant information ("intelligence") in the course of their routine business affairs. An ongoing police relationship with such partners, aimed at protecting targets, thus can be helpful.

In sum, we propose that no radical measures are needed by local police departments to deal with terrorist attacks. Rather, the development of an opportunity-reducing framework of local policing (a "problem-oriented" approach) serves the dual benefits of preventing and mitigating both crime and terrorism. We recognize that developing this opportunity-reducing framework is not a simple matter. Considerable training is needed for local police in understanding problem-oriented policing and situational crime prevention, in acquiring more specialized skills such as risk assessment and identification of vulnerable targets,[10] and in steps needed to develop partnerships with businesses and other organizations.[11] While this requires investment of valuable time and resources, the dual benefits of this approach make it an attractive and workable proposition.

NOTES

1. National Commission on Terrorism Attacks upon the United States. *The 9/11 Commission Report: Final Report of the National Commission on Terrorism Attacks upon the United States.* New York: Norton; 2004.
2. Newman GR, Clarke RV. *Policing Terrorism: An Executive's Guide.* Washington, DC: USDOJ Center for Problem-Oriented Policing; 2008. Brief 42.
3. National Security Preparedness Group. *Bipartisan Policy Center.* 10th Anniversary Report Card, Status on the 9/11 Commission Recommendations, 2011. Available at http://www .bipartisanpolicy.org/sites/default/files/911ReportCard.pdf

4. Finnegan W. The terrorism beat: How is the NYPD defending the city? New Yorker, July 25, 2005. pp 58–71.

5. Kelling GL, Bratton WJ. *Policing Terrorism.* Civic Bulletin 43, Manhattan Institute, New York, September 2006. Available at http://www.manhattan-institute.org/html/cb_43.htm, accessed June 21, 2012.

6. Braga A. *Problem-Oriented Policing and Crime Prevention.* 2nd ed. Monsey, NY: Criminal Justice Press; 2008; Goldstein H. *Problem-Oriented Policing.* New York: McGraw-Hill; 1990, and Philadelphia: Temple University Press; 1990. Available at http://www.popcenter.org/library/reading/, accessed June 21, 2012.

7. Clark RV, editor. *Situational Crime Prevention: Successful Case Studies.* 2nd ed. New York: Harrow and Heston; 1997.

8. Clarke RV, Newman GR. Police and the prevention of terrorism. Policing 2007;1(1):9–20.

9. Newman, Clarke, 2008; Clarke RV, Newman GR. *Outsmarting the Terrorists.* Westport, Connecticut and London: Praeger Security International; 2006.

10. Newman, Clarke, 2008; Boba R. Evil done. In: Freilich JD, Newman GR, editors. *Reducing Terrorism through Situational Crime Prevention.* Crime Prevention Studies, Volume 25. Boulder, CO: Lynne Rienner Publishers; 2009.

11. Scott MS, Goldstein H. Shifting and sharing responsibility for public safety problems. Response Guide No. 3, 2005. Available at http://www.popcenter.org/responses/responsibility/. Chamard S. Partnering with businesses to address public safety problems. Tool Guide No. 5, 2006. Available at http://www.popcenter.org/tools/partnering/, accessed June 21, 2012.

17

A MODEL CASE OF COUNTERTERRORISM: THWARTING A SUBWAY BOMBING

Joshua Sinai

THE PLOT TO BOMB THE NEW YORK SUBWAY

Najibullah Zazi's plot to bomb the New York City subway system in September 2009 was timed to coincide with the eighth anniversary of 9/11. Zazi and two high school friends planned to strap explosives to their bodies and head for the Grand Central and Times Square stations, where they would board packed trains at rush hour and blow themselves up.[1] Zazi's plot also involved links to Al Qaeda managers in Pakistan who directed the operation, as well as operatives in Britain and Norway who were in the process of carrying out their own attacks in those countries.

Had Zazi's plot succeeded, it would have been the most lethal terrorist attack in the US homeland since 9/11, possibly eclipsing the 2004 Madrid train bombings, which killed 191 people and wounded 1800, and the 2005 London bombings, which killed 52 and injured more than 700. Like Zazi's plot, the Madrid and London attacks followed the Al Qaeda "script" of coordinated and simultaneous attacks. Zazi's plot was thwarted by counterterrorism measures not only at the federal, state, and local levels but also transnationally in coordination with British, Canadian, and Pakistani authorities.

This particular case is highlighted because it represents a paradigm shift[2] in the types of questions to be asked along with what is to be observed, scrutinized, and interpreted. It stands as a challenge to counterterrorism agencies as they respond to increased numbers of terrorist threats from American residents with a nexus to Al Qaeda and its affiliates.

Zazi's Radicalization into Terrorism

Najibullah Zazi was born in Afghanistan's Paktia province, which borders Pakistan to the east. He was the middle child with two sisters and two brothers. His family is part of

Local Planning for Terror and Disaster: From Bioterrorism to Earthquakes, First Edition.
Edited by Leonard A. Cole and Nancy D. Connell.
© 2012 John Wiley & Sons, Inc. Published 2012 by John Wiley & Sons, Inc.

the Afghani Zazi clan, which is a subtribe of the Pashtuns. Zazis also reside in northwest Pakistan where, in the early 1990s, Najibullah's family found refuge from Afghanistan's internal violence.

Najibullah's father, Mohammed Wali Zazi, left Pakistan for America around 1992,[3] found employment as a taxi driver, and became a naturalized US citizen. This enabled him in 1999 to arrange for his family, including 14-year-old Najibullah, to settle in New York City, where they became legal residents. The family lived in a two-bedroom apartment in Flushing, Queens, in a mostly Afghani immigrant neighborhood.

9/11 and the destruction of the World Trade Center prompted dissension within the Afghani community in Queens. It led to a rift in the Afghani mosque where the Zazi family worshipped when Mohammad Sherzad, its imam, criticized the Taliban government that had shielded Al Qaeda in Afghanistan. The Zazi family left the mosque in opposition to Sherzad's anti-Taliban sermons.

At Flushing High School, Zazi was an indifferent student, more interested in playing basketball than studying. At age 16 he dropped out of school. A few years later he began working at his father's food cart near Wall Street. He also occasionally drove a cab and ran a coffee truck, altogether earning about $800 a month.[4] At the same time, he became increasingly devout, praying at the Masjid Hazrat Abu Bakr mosque where he did volunteer work as a janitor. His religious fervor prompted him to chastise friends for their interest in popular music, calling them "dishonest to your religion." After he began to wear an Afghan tunic, people took increasing notice. "Najib is completely different," a neighbor remarked. "He looks like a Taliban. He has a big beard. He's talking different."[5]

Zazi's radicalization was influenced by YouTube videos featuring Dr. Zakir Naik, an extremist Muslim televangelist, who preached an unorthodox Muslim theology that endorsed polygamy and harsh Islamic criminal law.[6] Naik's sermons spoke favorably of Osama bin Laden, stating that "If [bin Laden] is fighting enemies of Islam, I am for him."[7] Zazi became enchanted with Naik's preachings, which "may have given Zazi a mirror for his own confused feelings as he struggled to start a family and make ends meet," according to one analysis.[8]

In 2006, Zazi traveled to Pakistan for an arranged marriage to his 19-year-old cousin. He returned to the United States alone, but then went back to Pakistan each year for a brief stay with his wife.[9] While in New York he sought to earn money for his family, but he constantly fell into debt.[10] At one point, between April and August 2008, he opened 15 new credit card accounts.[11] Meanwhile, Zazi's wife had given birth to two children, and although indebted, he told friends he hoped to bring his family to the United States someday.[12]

Going Operational

During 2007 and 2008, Zazi became intensely angry over the US war in Afghanistan. Adis Medunjanin and Zarein Ahmedzay, two friends who had also attended Flushing High School, were similarly disaffected, and the three of them frequently discussed their opposition to America's actions. Like Zazi, Ahmedzay was born in Afghanistan, was raised in the United States, and eventually found work as a cabdriver. Medunjanin immigrated from Bosnia in 1994, was naturalized in 2002, and later attended Queens College, majoring in economics.

As explained by Zazi at his plea agreement, "they were bitter over the deaths of civilians in his homeland."[13] As they crossed the threshold from radicalization into violent action, they came up with a plan "to join the Taliban—to fight alongside the Taliban against the

United States" in Afghanistan.[14] Although exact details are not publicly known, it can be hypothesized that with Zazi's financial support from his credit cards "surplus," on August 28, 2008, the three friends flew from Newark International Airport via Switzerland and Qatar to Pakistan, with Peshawar their ultimate destination.[15]

Once in Peshawar, however, their plans changed. At Najibullah's father's trial in July 2011, Amanullah Zazi, Najibullah's cousin, claimed that while living with his parents in Pakistan in 2008, Amanullah introduced Najibullah to a Pakistani cleric who arranged for a meeting with Al Qaeda operatives. In September, the three were taken to Waziristan, an area with a strong Taliban presence, where they met Ferid Imam.[16] Imam hosted the three friends at his home and was key to arranging for their military training and religious indoctrination.[17] According to an account by Medunjanin, Imam "provided religious instructions on the rewards of fighting and dying together."[18]

Imam also showed them videos of Al Qaeda's most prominent attacks, including 9/11, Madrid, London, "and various other suicide or martyrdom operations."[19] Imam, now 30, had immigrated to Canada from East Africa when a child. He left for Pakistan's Peshawar in March 2007 to engage in terrorism with his friends Maiwand Yar and Muhannad al-Farekh. All three had grown up in Winnipeg and attended the University of Manitoba.[20] They had met at the university, but disappeared before completing their studies.[21] On March 15, 2011, Imam was indicted by US authorities for providing material support for Al Qaeda and aiding and abetting the Zazi group's training.[22] Terror conspiracy charges were also filed against him in Canada,[23] in a criminal investigation known as Project Darken.[24]

While Najibullah Zazi and his friends underwent military training at an Al Qaeda training camp, he received special bomb-making instruction because of his knowledge of the New York subway.[25] He took lengthy notes and e-mailed them to himself, so he could access them upon his return to the United States. Al Qaeda operatives then tasked them to return to the United States and conduct a suicide martyrdom operation on the subway system.

It is reported that Zazi met Bryant Neal Vinas at one of Al Qaeda's training camps. Vinas, an American from Long Island, New York, was captured in November 2008 by Pakistani authorities who turned him over to American law enforcement. A convert to Islamist extremism, he reportedly revealed extensive information to his American interrogators about Al Qaeda's network in Pakistan. It is possible that he tipped them off about Zazi's activities, since it was during this period purportedly that the CIA picked up on Zazi's presence in Peshawar. Al Qaeda was known to arrange meetings there with potential foreign operatives.[26]

At some point, Zazi met with a low-level Al Qaeda "fixer" or facilitator in Peshawar, with whom he communicated later in 2009 after returning to the United States. Zazi was seeking his advice on mixing chemicals.[27] Attesting to the international makeup of the terrorist networks, in 2008, the same facilitator had also met in Peshawar with two Pakistani students from Britain. They were among a group whose plans to blow up British shopping malls were later thwarted.[28]

While in Pakistan's Waziristan region, Zazi and his two associates also met high-ranking Al Qaeda figures, including Adnan Shukrijumah and Saleh al-Somali. At the time, al-Somali was Al Qaeda's chief of external operations, in charge of plotting attacks worldwide. He was killed in a US drone strike in 2009, but plots he had set in motion were continued after his death.[29]

Shukrijumah, a 35-year-old Saudi-born Al Qaeda veteran, was viewed as al-Somali's successor. He had grown up in the United States, but left just before the 9/11 attacks.[30]

Meanwhile, in 2008, Shukrijumah and Zazi's facilitator had joined in recruiting Zazi and his friends for the New York subway operation.[31]

By late January 2009, Zazi and his two cohorts had returned to the United States to put the plan into action.[32,33]

Operationalizing the Plot, and Its Preemption

Although Zazi could have been under some sort of US surveillance while in Pakistan, his plot actually may have been uncovered by Scotland Yard. In early 2009, British intelligence had intercepted an e-mail from a senior Al Qaeda operative in Pakistan, instructing Zazi on how to implement his attack.[34] Scotland Yard then notified the FBI, which heightened surveillance efforts, including listening to Zazi's phone conversations.

Within days of Zazi's return in January 2009, he moved to Aurora, Colorado, to live with his aunt and uncle. He began working for several airport shuttle companies, driving a van between Denver International Airport and downtown Denver. He proved highly industrious, routinely working 16- to 18-hour shifts. "He was a regular kind of guy, but he worked hard, and he wanted money," said Hicham Semmaml, a fellow driver. In June, however, his uncle evicted him from the apartment for not paying rent. His parents then moved to Aurora from New York, and the three took a residence together, with his father obtaining employment as a cabdriver.

Interestingly, after Zazi's return, Al Qaeda barely supported his efforts, apparently not advancing him enough cash to purchase hydrogen peroxide–based products for the bombs. As a result, to fund his operation Zazi again piled debt onto several credit cards. Unlike more "professional" terrorist operatives who attempt to conceal their finances from authorities, however, Zazi filed for bankruptcy in New York State in March 2009. By then his accumulated debts totaled $51,000. His bankruptcy was discharged (validated) in August 2009. Since Zazi apparently was under surveillance during this period—as a result of the tip off by Vinas to US authorities—it is possible that his "bankruptcy" may have raised an additional warning flag.

Beginning around June 2009, Zazi proceeded to the operational phase. He accessed his bomb-making notes and researched sites on the Internet to find ingredients for the explosives including hydrogen chloride, hydrogen peroxide, and acetone. He used his notes to rehearse construction of the detonators. He also took trips to New York, meeting with associates to discuss the plan, the attack's timing, and where to make the explosives.

In the course of its surveillance, the FBI discovered during the summer that Zazi and several cohorts were purchasing large quantities of hydrogen peroxide and acetone products at Denver beauty-supply stores. His associates reportedly used stolen credit cards to purchase the bomb-making materials.[35] At the Beauty Supply Warehouse on East Sixth Avenue, investigators found Zazi's image on security tapes, pushing a cart full of hydrogen peroxide–based items down the aisle. Additional supplies were purchased in early September. A later examination of Zazi's laptop indicated that while still in Denver he had searched the Internet for locations of a home-improvement store in Queens where he could purchase muriatic acid, a diluted version of hydrochloric acid.[36]

Hydrogen peroxide and acetone, which are readily available in beauty supply stores, are components of triacetone triperoxide (TATP; also known as acetone peroxide), a highly volatile material. TATP was used in the 2005 London subway bombings and by Richard Reid, the "shoe bomber" in his abortive 2003 attempt to blow up an airliner en route from Paris to Boston.

On August 28 and again on September 6 and 7, Zazi checked into the same Aurora motel suite, where he experimented in the room's kitchenette by heating and mixing the chemicals.[37]

During this period, as the FBI listened in on Zazi's telephone calls, he talked about "chemical mixtures and other things." Another intercept indicated that he had sent an e-mail message to his intermediary, known as "Ahmad," suggesting, in code, that the plot was nearing the attack stage.[38] "The wedding cake is ready," Zazi allegedly wrote.[39] About the same time, Zazi was trading e-mails with Ahmad and Naseer, who were part of a Manchester, England-based cell. In one message, Zazi e-mailed Ahmad that "the marriage is ready" shortly before he drove from Colorado to New York City carrying the bomb components.[40] In another text message, he wrote, the "wedding cake is ready," which may have been code to indicate the attack was imminent.[41] It is reported that in their e-mails, Al Qaeda operatives use references to weddings to disguise upcoming terrorist attacks.

On September 9, Zazi used his father's name to rent a car, which he proceeded to drive from Denver to New York City. In the car, there were detonators, explosives, and other materials necessary to build bombs. But he was followed by FBI agents.

In the afternoon of September 10, Zazi's car was stopped on the George Washington Bridge as he was about to enter New York City. He was told it was a random drug checkpoint (other cars were being stopped as well). Later, upon arriving in Queens, Zazi and several associates attempted to rent a U-Haul van, for possible transport of the operatives and their backpacks to the chosen subway stations.[42] Zazi spent the night at the residence of his childhood friend Naiz Khan, in Flushing, Queens. He intended to obtain and assemble the remaining components for a bomb over the weekend.

In a related development on September 10, two detectives from the intelligence division of the New York City Police Department (NYPD) interviewed Ahmad Wais Afzali, a Muslim cleric. The imam had been a police informant over the years, and the detectives asked him about Zazi and his associates. It is not known, however, if this interview had been coordinated with the FBI, whose agents had been trailing these suspects. Afzali admitted that he knew them and that Zazi had prayed at his Queens mosque. On September 11, Afzali called Zazi's father, who then spoke with Najibullah, telling him that "they" had shown Afzali his photos and photos of others. Zazi's father added, "So, before anything else, speak with [Afzali]. See if you need to go to [Afzali] or to make . . . yourself aware, hire an attorney." Afzali also called Zazi and told him that the authorities had asked him about "you guys." He also asked Zazi for the telephone number of one of the other men whose photos he had been shown, and set up a meeting with him.

On September 11, Zazi's rental car, which was parked near the Queens residence, was towed on a "purported" parking violation and was searched by FBI agents.[43] During a legally authorized search of Zazi's impounded car, the FBI found his laptop computer which contained images of multiple pages of handwritten notes. They included information on the manufacture and handling of explosives, detonators, components of a fusing system, all of which were necessary to produce an explosive charge.[44] The recipe for homemade explosives found on Zazi's computer would have produced bombs of the same size and type used in London, along with the requirement for backpacks and plastic containers—which were also used by the London bombers.[45]

The laptop also revealed that Zazi had researched baseball and football stadiums and sites used in the recent Fashion Week event in New York City.[46] In addition, it was reported that Zazi repeatedly had ridden the subway to the Grand Central and Wall Street stations, scouting for the best location to kill the maximum number of people. But after being tipped off by his father that he was under surveillance, he decided to abort the operation, and on September 12, he flew back to Denver.

Realizing that law enforcement was investigating him, Zazi and his associates began taking steps to dispose of the detonator explosives and other materials. In Denver, Zazi's father, Mohammed, ordered his family members to get rid of his son's bomb-making materials, which were stored in the garage, and lie to the FBI if they were questioned.[47] This was revealed by Amanullah Zazi, Najibullah's cousin, who testified at his father's trial.

On September 14, 2 days after Zazi had left New York, FBI investigators executed search warrants on several addresses in New York City, including where Zazi had stayed with his friends. In one of the apartments, they found 14 newly purchased backpacks, an electronic "black scale," and batteries containing Zazi's fingerprints.[48] The scale could be used to measure chemicals for hydrogen peroxide–based explosives, which the feds suspected was the purpose. An alert was then issued to American law enforcement officials to be on the lookout for hydrogen peroxide–based bombs.

The plan had been to assemble the remaining bomb components so that Zazi and his two associates could conduct coordinated suicide bombings on September 14 (the most likely date), 15, or 16. They intended to detonate four backpack bombs during rush hour on trains near New York's two busiest subway stations, the Grand Central and Times Square stations. The plan also called for boarding in the middle of packed trains on the 1, 2, 3, and/or 6 lines, with the aim of causing maximum casualties.

ARREST AND INCARCERATION

After returning to Denver, on September 16, Zazi and his attorney met with federal agents at the FBI office, where he was asked to submit DNA, hair, and handwriting samples.[49] During an 8-hour interview, he denied knowing anything about the handwritten pages on his hard drive and speculated that he may have accidentally downloaded them as part of a religious book that he had downloaded and later deleted. In subsequent interviews on September 17 and 18, however, he acknowledged receiving explosives and weapons training in the tribal areas of Pakistan.

On September 19, 2009, the FBI arrested Zazi and his father in Denver, while imam Ahmad Afzali was arrested in New York City. On September 21, Zazi was formally charged with making false statements involving international and domestic terrorism. Also on that day, the authorities arrested Zazi's father for conspiring to destroy evidence. He was released on $50,000 bond to his home in Aurora under house arrest, where he wore an electronic bracelet.

Ahmad Wais Afzali, Zazi's family's imam, was also arrested and charged with tipping off Najibullah. On March 4, 2010, Afzali pled guilty to a reduced charge in a plea deal and agreed to be deported to Saudi Arabia with his wife.

On January 7, 2010, knowing he was about to be arrested, Adis Medunjanin intentionally crashed his car on the Whitestone Expressway in Queens, with the intention of killing himself and other drivers.[50] He survived, called 911, and told the operator: "We love death."[51]

On February 22, 2010, Zazi pled guilty to conspiring to use weapons of mass destruction (explosives), conspiring to commit murder in a foreign country, and providing material support to a terrorist organization. His guilty plea was the result of a plea bargain with the prosecution.

In July 2011, Najibullah's uncle, Naqib Jaji pled guilty to his role in conspiring to destroy evidence against his nephew and promised to cooperate with the government as part of his plea agreement.[52]

IMPLICATONS FOR DOMESTIC TERRORISM

Najibullah Zazi's thwarted plot was part of a religiously based terrorist wave threatening America and its allies. Religious militants, a "bunch of guys," were joining with foreign terrorist groups in the global "jihad." Unlike the trained and well-funded foreign operatives on 9/11, Zazi and his associates were second-rate, "dispensable" figures directed to operate largely on their own. They were preempted due to the robust counterterrorism intelligence and law enforcement infrastructure established following 9/11. It now included close cooperation between the CIA and FBI and their security counterparts in other countries. In this case, that meant authorities in Britain, Canada, and Pakistan, all of whom helped uncover Zazi's plot in its early stages.

In tracking Zazi and his associates, the US government's counterterrorism agencies worked in an integrated manner to physically and electronically place them under surveillance and preempt their operation at the appropriately early moment. It should be pointed out that such surveillance is always conducted with prior court approval under the Foreign Intelligence Surveillance Act (FISA) when directed at threats such as terrorism. Unlike the pre-9/11 law enforcement-dominated paradigm, Zazi's conspiracy was allowed to proceed. Using proper civil liberties procedures, this enabled authorities to uncover and then apprehend the entire network.

One of the first times Zazi's name (and those of his associates) reportedly drew the attention of federal authorities was on August 20, 2008, when he took off from Newark to Pakistan. His name was entered into the Passenger Name Record database because of the potentially suspicious nature of his travel destination.[53] It surely came to their notice again when he returned to the United States on January 15, 2009.[54]

In terms of federal–local counterterrorism relations, the FBI worked closely with local law enforcement agencies. On September 9, 2009, when Zazi started driving his rental car from Denver to New York, the FBI obtained the cooperation of a Colorado State Patrol trooper to pull him over purportedly for speeding, but really to inquire about his destination. New York State troopers similarly cooperated when he reached the George Washington Bridge.[55] Some journalistic accounts claim that the NYPD's interview with imam Afzali tipped off Zazi that the authorities were on to him and that this could have thwarted potential prosecution.[56] But it seems as likely that the interview was coordinated by the FBI and NYPD to "smoke out" other potential associates including the imam and Zazi's father.

In the end, both men were arrested as coconspirators. Other members of Zazi's family were not only implicated but also testified for the government against the two men. In any case, Afzali's tipping off Zazi did not impair the investigation. Zazi apparently was left with insufficient time to destroy evidence that revealed how close he had come to build his bombs, place them in the backpacks, and lead his team to carry out the operation.

The FBI's role in the investigation also demonstrated how far the bureau had transformed into a robust domestic law enforcement and intelligence agency. Director Robert Mueller has indicated that the post-9/11 changes have been instrumental in enhancing the bureau's effectiveness. As he explained in congressional testimony, "the new approach prioritizes the collection and utilization of intelligence to develop a comprehensive threat picture, enabling strategic disruptions of terrorist networks before they act."[57]

Such focus on the overall threat picture reveals the value of sharing information between the FBI and state and local law enforcement agencies. The Zazi case was successfully resolved because, in the words of Senator Susan Collins, it "underscore[d] the incredible value of information sharing" and coordination among the law enforcement and intelligence components of these agencies.[58]

THE STRUCTURE OF COOPERATION AND COORDINATION

The FBI's offices in Denver and New York City, which are among the bureau's 56 field offices nationwide, were both involved in countering Zazi and his associates. They sifted through information, which then was disseminated locally and nationally. Recipients included the 100-odd Joint Terrorism Task Forces (JTTFs) nationwide and the National Joint Terrorism Task Force at FBI Headquarters, which comprises 40 member agencies.

As explained by Department of Homeland Security (DHS) Secretary Janet Napolitano in praising the Zazi investigation: "In that case, several FBI field offices and their JTTFs (including the New York JTTF) contributed to efforts in identifying Zazi, conducting surveillance of him, and arresting Zazi before he could execute his attack, while also identifying Zazi's associates."[59]

Another component of information sharing was conducted by the Colorado Information and Analysis Center (CIAC). The center provided analytic support to the Denver FBI and to the DHS, including personnel to assist the FBI in the investigation and support the field operations. According to the DHS website, "CIAC analysts also assisted in the review and analysis of the evidence obtained during the execution of the search and arrest warrants. CIAC leadership addressed media inquiries regarding the investigation, the threat to Colorado residents, and the threat to national security."[60]

The FBI has partnerships throughout the intelligence, law enforcement, and allied nations' security communities. Further, FBI legal attaches maintain liaison relationships with foreign counterparts, which also helped in Zazi's investigation. Other FBI divisions that were likely involved include the bureau's Terrorist Screening Center, established in 2003 and which maintains the government's Terrorist Watchlist (a database of individuals known or suspected of involvement in terrorist activity); the Terrorism Financing Operations Section, which tracks terrorist financing; and the Foreign Terrorist Tracking Task Force, also established in 2003, which provides information generated from government databases about individuals with a nexus to terrorism who may be operating in the United States. Once identified, leads on such terrorists and their associates are provided to appropriate law enforcement agencies to initiate investigatory and judicial action against them.

Although not designated a domestic counterterrorism agency, the National Security Agency (NSA) was also likely involved in the Zazi investigation. The NSA collects overseas signal intelligence, which consists of intercepting, processing, analyzing, and disseminating information derived from electronic communications and other signals employed by terrorist groups, their associates, and state sponsors. The NSA works closely with the FBI on cases involving Americans and foreign counterparts.

Also involved in helping to preempt Zazi's operation was the National Counterterrorism Center (NCTC).[61] The center was set up following the September 11, 2001 attacks, upon recognition that vital data on the pending strikes had not been shared between various intelligence agencies. Its mission is to synchronize the fight against terrorist threats within the United States and abroad, and to coordinate and share data on individuals connected to terrorism. The center's role in the Zazi investigation was recognized by President Barack Obama, who stated, "You [NCTC] stitched together the intelligence. You worked together, across organizations, as one team. And then—arrests in Denver and New York."[62]

Working with these and other intelligence and law enforcement partners enabled the FBI to deploy a discreet physical and electronic surveillance of Zazi once he returned to the United States in January 2009. While doing so, they ensured that he and his associates never posed, at any point, an immediate threat. This was also due to British intelligence's interception of his e-mails early on, sharing them with the FBI, as well as the FBI's phone taps and reading his e-mails before he began building bombs. From the time he left Denver for New York to carry out his operation, Zazi was under continuing surveillance by the FBI and local law enforcement agencies.

Good fortune also played a part. As in some other cases, such as Faisal Shahad's aborted attempt to bomb Times Square, Zazi was inept—constantly e-mailing his Al Qaeda contacts for advice on constructing the bombs. Moreover, publicly declaring personal bankruptcy and even buying bomb materials with maxed out and stolen credit cards must have sent red flags through the counterterrorism financial tracking system.

A final noteworthy success in thwarting Zazi's operation was the highly selective way in which information about the case was released to the public. There are still extensive information gaps about Zazi and his associates that may never be publicly revealed. Such restrictions protect possible ongoing investigations flowing from this and other cases, such as the one involving Vinas, who had met Zazi at an Al Qaeda training camp in Pakistan in 2008.

As mentioned earlier, Zazi's case study represents a paradigm shift in the nature and magnitude of terrorist threats to the United States. Unlike in the past, a primary threat now comes from American residents who have aligned with Al Qaeda and its affiliates and become radicalized into violent extremism. Even though their recruitment path was circuitous, Zazi and his associates have become the type of recruits Al Qaeda's operational planners are seeking: legal residents, acculturated in American society (even if hostile to its culture), and seemingly innocuous (although their activities would have raised red flags along the way).

CONCLUSION

The Zazi case shows that despite Al Qaeda's expulsion from Afghanistan after 9/11, and despite attacks on its operatives in Pakistan's lawless regions, the organization could still project terrorist violence into the United States. The targeted killings in 2011 of Osama bin Laden and other leading Qaeda figures dented the organization's command structure. But few believe that its many adherents have become less fervent.

Doubtless, foreign terrorists will still attempt to enter the United States to conduct operations against the homeland. But their efforts have become more difficult to execute because of the comprehensive antiterrorism infrastructures instituted after 9/11. As a result, the United States now faces an elevated threat from homegrown terrorism—a legal resident who is motivated, connected, and capable of executing an attack, even if less catastrophic than the one on 9/11.

While the Zazi plot heightened awareness about homegrown terrorists, it also exemplified the connection to operational planners in the multinational network of Al Qaeda and its affiliates. The network stretches from Pakistan's Taliban-dominated regions where Qaeda operatives have found sanctuary to Al Qaeda in the Arabian Peninsula (AQAP), which includes strongholds in Yemen. The vitality of these two centers of activity was demonstrated by two other organized efforts.

On Christmas Day in 2009, Umar Abdulmutallab, trained and directed by AQAP, sought unsuccessfully to blow up a Northwest flight bound for Detroit. Eight months later, in May 2010, Taliban-trained and directed Faisal Shahzad tried to detonate his explosives-laden van in Times Square. Had both efforts not failed in their execution phase, they would have caused catastrophic damage.

These and similar plots demonstrate that Al Qaeda groups have made a strategic decision not to solely pursue attacks on the scale of 9/11. Rather they are aiming at more modest tactical successes involving "only" tens or hundreds of casualties. As explained by former FBI agent Jack Cloonan: "They want to see bodies, blood sprayed all over the place. They want to punish us. . . . It accomplishes a number of things aside from body count. It reaffirms that they are alive and well."[63]

NOTES

1. Marzulli J. Zazi, Al Qaeda pals planned rush-hour attack on Grand Central, Times Square subway stations. Daily News, April 12, 2010.

2. This formulation is based on Kuhn T. *The Structure of Scientific Revolution*. Chicago: University of Chicago Press; 1966.

3. Burke K et al. How the feds caught Najibullah Zazi, pieced together the 9/11 terror plot. Daily News, September 27, 2009.

4. Meek JG et al. A dozen on constant watch including Najibullah Zazi in FBI's terrorist probe. Daily News, September 18, 2009. Burke K et al. How the feds caught Najibullah Zazi, pieced together the 9/11 terror plot.

5. Baker A. An enemy within: The making of Najibullah Zazi. Time, October 1, 2009.

6. Ibid.

7. Ibid.

8. Ibid.

9. Cedars D. Zazi questioned by FBI for hours. The DenverChannel.com, September 17, 2009.

10. Baker A. An enemy within: The making of Najibullah Zazi.

11. Meek JG et al. A dozen on constant watch including Najibullah Zazi in FBI's terrorist probe.

12. Ibid.

13. Susman T, Serrano RA. Guilty plea in New York terrorism case. Los Angeles Times, February 23, 2010.

14. Ibid.

15. United States of America v. Najibullah Zazi, Criminal Complaint, United States District Court for the District of Colorado, September 19, 2009. p 2.

16. Hays T. Father of NYC subway bomb plotter is convicted. Associated Press, July 22, 2011, Denverpost.com. Marzulli J. Father of would-be terrorist Najibullah Zazi hid bomb materials, relative testifies. Daily News, July 18, 2011.

17. Marzulli J. Ferid Imam provided terror training to three high school friends trying to join the Taliban: Feds. Daily News, March 15, 2011.

18. Ibid.

19. Ibid.

20. Ferid Imam, Canadian fugitive charged in NYC subway plot. Associated Press, March 15, 2011.

21. Terror charges laid against former Winnipeggers. CBC News, March 15, 2011.

22. Richey W. Canadian charged with helping Najibullah Zazi in New York bomb plot. Christian Science Monitor, March 15, 2011.

23. Ibid.

24. RCMP lay terrorism-related charges. Royal Canadian Mounted Police, March 15, 2011. Available at http://www.rcmp-grc.gc.ca/news-nouvelles/2011/03-15-darken-eng.htm, accessed June 31, 2012.

25. Marzulli J. Zazi, Al Qaeda pals planned rush-hour attack on Grand Central, Times Square subway stations. Daily News, April 12, 2010.

26. Ross B et al. FBI arrests three men in terror plot that targeted New York. ABC News, September 20, 2009. Available at http://abcnews.go.com/Blotter/men-arrested-fbi-nyc-terror-plot/story?id=8618732, accessed June 31, 2012.

27. Gardham D. New York subway plot was five days from success. The Telegraph, December 14, 2010. Available at http://www.telegraph.co.uk/news/worldnews/northamerica/usa/8202709/New-York-subway-plot-was-five-days-from-success.html, accessed June 31, 2012.

28. Ibid.

29. MacDougall I et al. Norway bomb plot connected to U.S., British threats. Associated Press, July 9, 2010.

30. Meek JG. Subway plotter Najibullah Zazi met with key al Qada player Adnan Shukrijumah, feds believe. Daily News, July 1, 2010.

31. Hays T, Apuzzo M. New York subway bomb plot linked to British cell. Associated Press, July 8, 2010.

32. Ross B et al. FBI arrests three men in terror plot that targeted New York.

33. United States of America v. Adis Medunjanin, Abid Naseer, Adnan el Shukrijumah, Tariq ur Rehman, and FNU LNU, p 8.

34. British spies help prevent al Qaeda-inspired attack on New York subway. The Telegraph, November 9, 2009.

35. Finley B, Cardona F. FBI: Terror suspect cooked bomb materials in Aurora. The Denver Post, September 24, 2009.

36. Ibid.

37. Ibid.

38. Gardham D. Manchester bomb plot students were planning co-ordinated attacks in New York and Scandinavia.

39. Ross B et al. FBI arrests three men in terror plot that targeted New York.

40. Hays T, Apuzzo M. Najibullah Zazi's Al-Qaeda bomb plot likely part of larger operation [Associated Press]. The Herald Sun, July 7, 2010.

41. Ross B et al. FBI arrests three men in terror plot that targeted New York.

42. Fletcher D. Terrorism suspect Najibullah Zazi. Time, September 22, 2009.

43. United States of America v. Mohammed Wali Zazi, Criminal Complaint, United States District Court for the District of Colorado, September 19, 2009. p 6.

44. Ibid.

45. Ross B et al. FBI arrests three men in terror plot that targeted New York.

46. Ibid.

47. Marzulli J. Father of would-be terrorist Najibullah Zazi hid bomb materials, relative testifies.

48. Meek JG et al. Feds unsure if arrest of Najibullah Zazi and two others has foiled al-Qaeda terror plot. Daily News, September 20, 2009. Available at http://articles.nydailynews.com/2009-09-20/news/17933259_1_najibullah-zazi-al-qaeda-fbi, accessed June 21, 2012.

49. Najibullah Zazi denies any ties to terrorist groups. TheDenverChannel.com, October 2, 2009.

50. "United States of America Against Adis Medunjanin, Abid Naseer, Adnan el Shukrijumah, Tariq ur Rehman, and FNU LNU," p. 9.

51. Ibid, p 8.

52. Reiss A. Trial begins for father of NYC bomb plotter accused of misleading FBI. CNN, July 18, 2011.

53. Rrausnitz Z. DHS defends passenger data collection amid E.U. privacy concerns, fiercehomelandsecurity.com, October 5, 2011.

54. Burke K et al. How the feds caught Najibullah Zazi, pieced together the 9/11 terror plot.

55. Burnett S. Colorado Zazi's coded e-mail started agencies plan to stop N.Y. subway attack. The Denver Post, October 2, 2011.

56. Burke K et al. How the feds caught Najibullah Zazi, pieced together the 9/11 terror plot.

57. Robert S. Mueller. "Ten Years After 9/11: Are We Safer?" Statement Before the Senate Committee on Homeland Security and Governmental Affairs, September 13, 2011. Available at http://www.fbi.gov/news/testimony/ten-years-after-9-11-are-we-safer, accessed June 21, 2012.

58. Statement of Ranking Member, Susan M. Collins, "Information Sharing in the Era of Wikileaks: Balancing Security and Collaboration," March 10, 2011. Available at http://www.fas.org/irp/congress/2011_hr/031011collins.pd, accessed June 21, 2012.

59. Testimony of Secretary Janet Napolitano before the United States House of Representatives Committee on Homeland Security, "Understanding the Homeland Threat Landscape—Considerations for the 112th Congress," February 9, 2011. Available at http://www.dhs.gov/ynews/testimony/testimony_1297263844607.shtm, accessed June 21, 2012.

60. Department of Homeland Security, "Fusion Center Success Stories: Fusion Center Supports Zazi Investigation," September 2009. Available at http://www.dhs.gov/files/programs/gc_1296488620700.shtm, accessed June 21, 2012.

61. "Obama vows to flush out Al-Qaeda," Agence France Presse, October 6, 2009.

62. Remarks by President Obama at the National Counterterrorism Center (NCTC), October 6, 2009. Available at http://www.dni.gov/speeches/20091006_speech.pdf, accessed June 21, 2012.

63. Thomas P. NYC subway plot: Dangerous new phase in threat by Al Qaeda. ABC News.com, February 23, 2010. Available at http://abcnews.go.com/GMA/najibullah-zazi-nyc-subway-plot-al-qaeda-threat/story?id=9917485, accessed June 21, 2012.

18

THE NEWEST SECURITY THREAT: CYBER-CONFLICT

Panayotis A. Yannakogeorgos

The predominant purpose of cyberspace is to serve human operators and create effects in the physical world. In June 2010, the Stuxnet computer worm infected the digital control software of the Iranian nuclear program creating physical damage to the cascading centrifuges churning nuclear material. The world at large may not have been familiar with physical effects of cyber events on infrastructure before Stuxnet, but since 2010 the vulnerabilities that exist within cyber infrastructures have made it apparent that code can cause machines to destroy themselves and cause their operators to mistrust the information residing on them.

Horrid possibilities emerge when thinking about the exploitation of cyber events during a natural disaster, and its impact on the medical community and first responders. Indeed, the US government considered it a national security priority to protect health infrastructures from cyber-based terrorist events that could "cause catastrophic health effects or mass casualties comparable to those from the use of a weapon of mass destruction."[1] The use of cyberspace is the most recently recognized threat that can exploit vulnerabilities within the healthcare communities' digital infrastructure.

To date, there has been no large-scale, nationally significant cyber event during a disaster or terrorist event from which to judge the effects and costs of a cyber attack on the medical community. However, the impact of a cyber event could be massive. A Department of Homeland Security (DHS) report concluded that during a natural disaster event:

> Cyber terrorism may be the most likely and potentially dangerous attack that could be mounted easily—locally or internationally—during all phases of the storm. . . . A cyber attack could produce erroneous information on the storm to confuse the population, hamper critical infrastructures during the storm thereby increasing instability and response efforts. Denial of service attacks on infrastructures such as the 911 [phone-call] system could be widespread and impact communication between first responders and the public.[2]

Local Planning for Terror and Disaster: From Bioterrorism to Earthquakes, First Edition.
Edited by Leonard A. Cole and Nancy D. Connell.
© 2012 John Wiley & Sons, Inc. Published 2012 by John Wiley & Sons, Inc.

Information and communication technology is essential for the effective command and control of first responders to a natural disaster. This chapter begins with a case analysis giving an overview of the US military response to Hurricane Katrina and the interaction among federal and local first responders emphasizing information and communication technology. The case analysis will include lessons learned from certain events occurring in the aftermath of the hurricane that had an impact on cyber assets. Next, a "primer on cyber" will provide background material on the nature and components of cyberspace and the role of cyberspace in the medical field. Then the actors involved in cyber events are described (facilitators, defenders, targets), followed by current US government organization of security. The chapter concludes with a discussion of challenges in maintaining cybersecurity.

Awareness of the vulnerabilities and threats in the cyber domain is perhaps the first line of defense again cyber attacks. The intent of this chapter is to both inform and raise awareness within the medical and responder communities of vulnerabilities and protection.

THE US AIR FORCE AND THE HURRICANE KATRINA KILL ZONE

On August 29, 2005, Hurricane Katrina made landfall just east of New Orleans, Louisiana. Damage from the hurricane and floodwaters besieged municipal and state disaster response capabilities. Department of Defense (DOD) teams provided large-scale relief efforts in response. The Pentagon's response to Hurricane Katrina was the largest deployment of military forces for a civil-support mission in US history.[3] The Air Force (AF) took a large role in military operations. The AF aided in a matchless effort to evacuate endangered populations and provide humanitarian relief. The AF team included active duty, Air National Guard, Air Force Reserve, and Civil Air Patrol volunteers from the Air Force Auxiliary. Airmen efforts included, but were not limited to, search and rescue, logistic planning, and aeromedical evacuation.

The aeromedical rescue effort was expanded partly due to the exemption of hospital and nursing homes in the New Orleans mandatory evacuation order. Hospitals accounted for 1749 high-maintenance patients and 7645 staff and guest requiring evacuation.[4] AF assets were used to evacuate patients remaining in the hospitals and nursing homes. Hospitals had emergency procedures for storms, but the massive flooding demolished these well-intentioned plans. The floodwaters disabled backup generators located in basements and the damaged electrical systems could not sustain basic life support functions. In addition to not having electrical lifesaving equipment, all electronic data were inaccessible, including patient records regarding medications, allergens, drug interaction warnings, and medical histories.

Challenges in communication pathways between military responders and civilian entities came to light during the rescue mission. The means to communicate effectively throughout an enterprise were not available. With the loss of electrical power throughout the area, communication was obstructed. Landlines, cellular towers, and radio repeaters were deemed inoperable. With this loss of capability, emergency operation centers were isolated from information sources.[5] Broken communication channels impeded communication between AF and civilian entities.

The data pertaining to the situation were available, and very robust but the paths across service, joint and interagency, were absent. This resulted in a fragmented common situational awareness. Therefore, unity of effort between AF and civilian entities was impeded, which consequently delayed rescue efforts. There was no single common operating picture for all interested parties to work with. Finally, the Internet, which has become a widely used basis for communication, was down in the area. Thus, Internet-based software for processing patient movements was not accessible to civilian medical personnel due to the destroyed infrastructure.

Communication between parties is critical for effective response. First, communication is required to establish initial situational assessment (location of survivors, condition of infrastructure, participating rescue and recovery organizations, security conditions). Second, communication ensures that efforts are not duplicated among participating organizations. Contact between entities must be adaptable to changing conditions, such as further deterioration of infrastructure, locating additional survivors, changing roster of rescue and recovery organizations, or changing security situations. There are patterns of interactions that can be studied, such as who needs to interact, how they interact, and what types of transactions occur during the interaction. The objective is to ensure that all those involved across functions, organizations, and echelons understand the mission and are working in concert toward the desired end.[6]

The Katrina operations were conducted in the continental United States, and they should have carried a "home field advantage." The involved communities were familiar and friendly, not foreign or adversarial. The multiservice government and civilian entities should have been able to work together seamlessly. Operating independently influences decisions that affect all parties and can delay critical capabilities. For example, the use of military unmanned aerial vehicles (UAVs) for support to civil authorities (for surveillance and reconnaissance operations) was banned by the Federal Aviation Administration. This decision illustrates the general lack of integration of capability and capacity of military assets. Integration of capabilities between parties can be accomplished with proper planning and exercises. Proper planning with current capabilities could have prevented delays.

PRIMER ON CYBER

Cyberspace has played a significant role in the medical community for the final half of the last century, and the first decade of the twenty-first century. The public health system relies on digital technologies to improve treatment via efficient emergency response communications, digital patient and pharmaceutical records, and advanced equipment such as magnetic resonance imaging (MRI) machines. Hospitals rely on supervisory control and data acquisition systems to allow for the remote control and monitoring of elevators, energy management, and heating, ventilation, and air conditioning (HVAC) system control. Whether it is a 9/11 emergency call that is communicated via digital devices networked via the electromagnetic spectrum, or a patient's medical record that is digitally stored, cyberspace is being used to create effects in the physical world.

As noted earlier, during Katrina, floodwaters and natural events damaged the digital infrastructure, not malicious individuals conducting a cyber attack. While there is no argument against the utility of man-made elements of cyberspace, focusing too much on the technology creates conceptual hazards that clouds policy discussions.[7] Focusing on technology rather than the characteristics that make up the entirety of the cyber environment creates the impression that this domain is not connected with the real world. Recognizing that cyberspace is more than just a virtual environment is beneficial for the healthcare community. Refining the conceptualization of cyberspace will allow for its demystification,

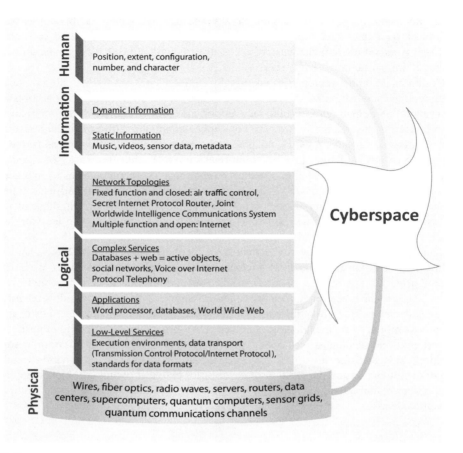

FIGURE 18.1 Characteristic-Based Model of Cyberspace. (For a color version of this figure, see the color plate section.)

and a closer alignment within the physical world.[8] This requires looking at cyberspace as a complex ecosystem composed of a number of elements: human operators ranging from the casual Internet user to the terrorist; the actual information that is stored, transmitted, and transformed; the computer code and protocols; and the physical elements on which the logical elements reside.[9] Figure 18.1 is a graphic representation of a holistic view of cyberspace.

The human and physical aspects are just as important as the logical elements of cyberspace. Data and information are not transported in a virtual ether divorced from the laws of physics, space and time. Rather, data and information travel through physical infrastructures, such as undersea cables, and reside on digital storage devices operated by people whose behaviors are organized within policy and regulatory environments such as the Health Insurance Portability and Accountability Act. Software and hardware are developed by companies whose products may be the root of vulnerabilities due to poor system design and implementation or vulnerabilities in the hardware. Currently, the private sector, which is responsible for operating a large portion of cyberspace, is not regulated. Thus, system designers are not held responsible for vulnerabilities they introduce into critical infrastructure. Some cyber operations also affect the cognition of people who use the data residing on information systems to make decisions. If a first responder cannot trust the data, there is a negative impact on his ability to provide swift medical attention where it is needed.

Computer networks are dependent on the use of internationally standardized communication protocols, known as the Transmission Control Protocol and Internet Protocol (TCP/IP), to send and receive data-packets and information.[10] TCP/IP allows for the flow of data-packets and information across computer networks, including the Internet. The way machines identify each other on the Internet is through Internet Protocol (IP) and Media Access Control (MAC) addresses. IP was not intended to function as the backbone of the global project that became the Internet. Designed and deployed for military and research purposes in the late 1960s, the ability to track and trace user behavior in a highly untrustworthy computing environment was not embedded into the design of communications protocols.[11]

Thus, this foundational protocol on which the Internet runs is too weak to provide reliable security mechanisms.[12] According to Internet expert Tom Leighton, the domain name system and the IP address system are plagued by flaws that "imperil more than individuals and commercial institutions. Secure installations in the government and military can be compromised" as well.[13] Consequently, the current flaws in the network architecture of the Internet are a result of relying on the protocols that were built 35 years ago when the Internet was not a global entity, but a closed research network. As the Internet became global, there was no shift to create stronger security mechanisms.

To understand better the functioning of TCP/IP, a brief description of how information is sent across networks is necessary. Data-packets are the basic units of network traffic. They are the standard way of dividing information into smaller units when sending information over a network. A significant component of the computer networks is the IP header, which contains information pertaining to the source and destination addresses. Machines require these strings of numbers to connect with other computers on the Internet or other networks.[14] All networked hardware must have a valid IP and MAC address to function on a network. Data-packets are recreated by the receiving machine based on information within a header of each packet that tells the receiving computer how to recreate the information from the packet data. Without international standards, such as TCP/IP, there would be no assurance that packets could be read by a receiving machine.[15]

All of the elements of cyberspace in the model (Fig. 18.1) have vulnerabilities that can be exploited by malicious actors during a natural disaster. Data transport protocols are vulnerable to "spoofing" attacks that is at the root of the attribution problem: this method could mask the attacker's identity. Not knowing who is attacking makes it difficult to know how to respond. While ability for malicious actors to disguise offline human and machine identity will be reduced with the deployment of the next generation Internet networking protocols, known as Internet Protocol version 6 (IPv6), these technical improvements will not prevent a determined adversary from causing damage. Identifying information can be found in other layers of information that could aid in attributing a cyber attack to actors. But US government resources at state, local, and tribal levels would be required to tackle this issue and would quickly be exhausted in a natural disaster setting. National response plans are covered in greater detail below, and international cooperation will be required as well to reduce the levels of ambiguity that currently exist in cyberspace.[16]

Attack Agents

Attack agents can be states, sub-state actors such as Chinese privateer hackers or Romanian computer criminals, regional/global organizations such as the Russian Business Network, Ad-hoc networks such as the LulzSec hacking group, or malicious individuals such as an unreformed Kevin Mitnick, and the proverbial threatening insider.

While all forms of exploitation of vulnerabilities within information systems pose a threat, not all attack agents are capable of producing an effect in cyberspace that would pose a threat to national security. A complex, advanced level attack of national significance accompanied by stalled attribution would require

- Expert level programming and cryptographic skills
- Detailed knowledge of industrial control systems
- Mastery of multiple open and closed operating systems
- Detailed knowledge of telecommunications and legal regimes

Motivating factors for an attack are also important when gauging the intention of the attack agent, such as identity theft, espionage, zombie propagation, extortion, sabotage, or widespread destruction. The first four of these often indicate economic incentives in which the perpetrator of an attack judged that an investment of time and other resources would bring about a higher payoff. The goals and objectives of an attack include the corruption, fabrication, destruction, and disclosure or discovery of information. System subversion or disruption might be additional or collateral goals. Cyber events occur through system or protocol compromise, resource exhaustion, hardware failure, or software crashes. The techniques used to achieve the objectives include the targeted exploitation of system, social, or protocol vulnerability. Other techniques include overload of network or system resources and the autonomous self-propagation of malware.

Facilitators

Attackers will aim to mask their identity to avoid prosecution, especially those that are motivated by economic gains through criminal exploits or intent on widespread destruction. As noted earlier, a sophisticated attack agent will likely have a sound knowledge of telecommunications and legal jurisdictions that would allow the routing of an attack through countries lacking legal authority or expertise to prosecute cyber criminals, or minimum standards in data retention. Facilitators also include countries lacking networks and processes of international cooperation.

Facilitators could also include unwitting computer users whose equipment has been infected with malicious code allowing remote control at a time of the malicious agent's choosing. The facilitating condition for this is lack of awareness, training, and education about cyber threats.

Other facilitating actions include reluctance or outright refusal to cooperate in an investigation. All of these conditions result in the sustaining of havens from which malicious cyber agents will operate knowing that their identities can be hidden from law enforcement and others.

Service providers and software companies themselves can serve as attack facilitators. Internet Service Providers (ISPs) in the west are often reluctant to monitor network traffic for malicious content. A recent push by the White House on ISP monitoring for copyright infringements is evidence that ISPs can be used to limit the environments through which an attack can pass.[17] Software companies and suppliers of hardware are facilitators in that they produce the physical and logical components on which cyberspace exists. Hardware supply chains have been found infected with bugs during manufacturing in plants outside of the US software companies. Concerned with the next quarterly financial report, the companies may push products onto the market before they can be fully tested for security.

In fact, many commercial programmers are not trained to write secure code. Many mission-critical information systems are still coded in security weak language. Insecure software is released without automating secure coding practices prior to market launch or choosing different coding languages. Once on the market, when vulnerabilities are discovered, "patches" are issued to secure the computer from potential attack agents. However, most users do not update their software with the latest path or antivirus definitions. All the above factors facilitate conditions for cyber attacks by creating an ecosystem that highly favors the attacker and leaves the defenders in a post-attack reactive posture.

Defenders

Defenders of cyberspace include ISPs, law enforcement, corporate security branches, national computer emergency response teams, international efforts such as the International Telecommunications Union's International Multilateral Partnership Against Cyber Threats (IMPACT), and the individual computer user. Not all countries take a similar stance on cyber defense. Global cybersecurity is hindered by lack of an organizing defense through cybersecurity action plans at the national level. Such plans implement the technological, managerial, organizational, legal and human competencies into national security strategies to enhance defense.[18] Criminals, privateer-hacker networks, and information warriors exploit countries lacking these structures in order to launch cyber attacks of national and global significance. Indeed, the vitality of American social, economic, and governmental institutions are at great risk from cyber vulnerabilities existing in less developed countries.[19] Reducing domestic vulnerabilities and threats from cyber attack hinges on the US policy community supporting norms of behavior among states, enforceable at the national level, to secure the cyber commons.[20]

Targets and Effects

Humans are targets for social engineering campaigns that aim to exploit trust relationships among computer users. The recent attack on RSA, the security division of an information technology management company called EMC, is instructive. ("EMC" derives from Einstein's famous equation: $E=MC^2$, and "RSA" is a non-descript acronym perhaps deriving from the words risk and security). When RSA employees interacted with a malicious e-mail message they unknowingly gave the attack agents access to RSA's networks.[21] Other targets include hospital HVAC systems, global positioning systems (GPSs), and data storage points. A cyber-related event will produce effects that depend on the perpetrator's motivation for launching an attack. Cyber events can be either discrete and finite, or advanced and persistent. An example of a discrete and finite event is an operation aimed at degrading a GPS by shifting mapping vectors and distorting positioning. Advanced persistent threats are those linked with espionage and criminal activities that aim to gain as much information about the functioning of a system as possible. Within a medical context, these threats could exploit patient record databases to steal social security numbers and other personal identity information, which in the long term could be used to forge identity records and open fraudulent financial accounts.

Effects are observed either as the result of a cyber disruption within a service or the cascading disruption resulting from the system's dependency on another service directly affected. The services affected include the sectors of energy, telecommunications, finance, water supply, healthcare, transportation, law enforcement, fire and emergency response, government administration, shipping, agriculture, commercial facilities, and critical manufacturing. The impact of an event could be detrimental to the economy, population, or government. Motivating factors also play a role in the response. The severity of the effect of a cyber attack will determine whether a response will cross over the defense threshold. Unlike criminal attacks, which usually involve widespread and indiscriminate targeting to obtain maximum profit from their victims, cyberweapons will be more focused on their intended target.

Cyberspace is not necessarily the primary factor in determining vulnerability of systems. Criminals and cyber warriors will target institutions regardless of whether there is a way to do it via cyberspace. Many argue that the cost of entry is low in cyberspace because it is relatively simpler to launch the digital version of a bank robbery, meddle with a hospital's HVAC system, or release the floodgates on a damn. It is true that one requires significantly fewer resources to conduct a cyber attack, yet the reason for this has less to do with the nature of the domain than with poor product development, design, and implementation. Software developers, hardware manufacturers, and network providers face no liability or responsibility for the systems they produce or operate. As a result they have no legally based incentive to deliver secure products to the marketplace. This risk will really begin to manifest as cloud computing takes hold and resulting breaches will destroy multiple points within the healthcare establishment rather than in a single hospital.[22] (Cloud computing delivers computing information as a service, comparable to the way electricity is delivered via a grid.)

Following are the kinds of effects that a cyber attack could have on first responders during terrorist events:

- Degrade GPS by changing time stamps.
- Attack information systems related to supply and logistics and manipulate the records so that supplies and hospitals do not correlate. This could result in groups of people being sent to the wrong hospital.
- Attack 911 systems so that people are evacuated from areas that are not in danger.
- Target pharmaceutical manufacturing hubs so that chemical imbalances are introduced in the supply.
- Target data storage points for information later used to exploit people.
- Attack water treatment plants to release dangerous volumes of chlorine while making levels appear normal to operators.
- Target choke points such as drawbridges, to prevent rescue operations or evacuations.
- Create "spoof" 9/11 calls.
- Attack blood test data banks to manipulate data so that machines displaying results in clinical tests would indicate a pathogen, such as anthrax, is present in blood work. Two successful cases in one area could result in an unwarranted quarantine, which would prevent the evacuation of civilians and bringing in of emergency responders.
- Attack medical patient databases leading to inappropriate treatments, such as amputations.

- Send false information over messaging systems, for example, instructing people to shoot at emergency responders.
- Shift GPS grid maps by a few blocks to confuse first responders, or lure them into a booby-trapped area.

US GOVERNMENT ORGANIZATION

A brief history of the federal policies on cyber is indicative of challenges to organizing the military-civil sector. Before a natural disaster, or after a significant hard strike by a terrorist organization, cyber escorts need to be in place patrolling the cyber domain in an effort to determine whether terrorists or other actors are attempting to amplify the effects of a natural or man-made disaster with a cyber strike. This surveillance system is the end state toward which efforts should be leading.

An effective response to a cyber incident of national significance requires a quick and well-coordinated response by all relevant actors. In June 2008, the Government Accounting Office (GAO) considered the integration of cybersecurity offices and centers within the US government as critical for prompt government action, including integration of Department of Homeland Security/Computer Emergency Readiness Teams (DHS/US-CERT) and the National Coordinating Center for Telecommunications (NCC).[23] The NCC is jointly run by the Federal government and the telecommunications industry to coordinate the exchange of information among the participants regarding vulnerability, threat, and intrusion as they affect the telecommunications infrastructure.[24] The GAO notes that DHS/US-CERT and the NCC have overlapping missions in the areas of

- Developing and disseminating warnings, advisories, and other urgent notifications
- Evaluating the scope of an event
- Facilitating information sharing
- Deploying response teams during an event
- Integrating cyber, communications, and emergency response exercises into operational plans and participation
- Managing relationships with others, such as industry partners[25]

Partially heeding GAO advice, the Department of Homeland Security moved the NCC to offices adjacent to US-CERT in November 2007. The two centers now have adjoining office space and use common software to "identify and share physical, telecommunications, and cyber information related to performing their missions."[26] However, they were not merged into one joint operations center as recommended by GAO until November 2009. Other bodies within Homeland Security also duplicate several functions, including the National Cyber Security Division (NCSD) and the National Communication System (NCS). Both were identified by the GAO and the Obama Administration's *Cyberspace Policy Review* as two centers requiring integration.

An organizational merging of the functions of the NCSD and NCS did not occur immediately due to competing priorities and differing interpretations of their responsibilities.[27] A Homeland Security-commissioned expert task force warned at the time "that without an organizationally integrated center, the department will not have a comprehensive operating picture of the nation's cyber and communications infrastructure and thus not be able to effectively implement activities necessary to prepare, protect, respond, and recover this infrastructure."[28]

In October 2009, the National Cybersecurity and Communications Integration Center (NCCIC) became operational, addressing the concerns discussed above and implementing key elements of various White House initiatives including recommendations by the *Cyberspace Policy Review*. Thus, after many years of moving to merge the US-CERT and NCC, the Department of Homeland Security is now leading a 24-hour watch and warning center that aims to protect critical components of the cyber infrastructure.[29]

CHALLENGES IN SURVEILLANCE

While many cybersecurity concerns are now acknowledged, limited resources within responsible agencies put a ceiling on the number that are fully being addressed. DHS and various presidential directives still place primary emphasis on securing federal computer systems, but not enough attention is being given to the fusion of multiple centers. Overall, languid implementation of some cybersecurity strategies has resulted in continued vulnerability in the nation's computing environment. Turf conflicts between some agencies have exacerbated the situation. Thus, Rod Beckstrom, former director of DHS's National Cybersecurity Center, resigned from his position in 2009 unhappy with the National Security Agency's control of DHS cyber efforts and its dominant role in cybersecurity.[30] To rectify the mistrust between the two agencies, in 2010, DHS and NSA signed a Memorandum of Agreement that enhanced their cooperation on matters affecting the intersection of homeland and cybersecurity.[31] Similar cooperative frameworks will enhance protection for the medical and other critical information infrastructures of the United States.

CONCLUSION

There is one indispensable ingredient to securing medical cyberspace that was not discussed in this chapter. That is the user, whether health worker, responder, or other member of the public. The possibilities described above are predicated on what we know about cyber vulnerabilities and threats in cyberspace. What we do not know is how innovative threat actors will exploit public health workers and others who are unaware of the dangers of cyberspace, and allow breaches to take place. Lessons from Hurricane Katrina are indicative of the physical vulnerabilities to information systems that can affect operations.

The impact of a natural disaster may be magnified by a cyber attack that aims to destroy infrastructure, deny access to information, or degrade first responder operations. The considerations here demonstrate that we cannot seek national security in cyberspace within silos. Private-public partnerships are required. A society-wide approach for cybersecurity is required, and the medical community has much to contribute to protect public health services. Time is short before malicious actors in cyberspace reach a stage when they track a hurricane and launch cyber attacks to degrade first responder efforts.

NOTES

1. Homeland Security Presidential Directive 7: Critical Infrastructure Identification, Prioritization, and Protection, 2003.

2. Terrorists Could Exploit Hurricane in U.S., Report Warns. Global Security Newswire, April 25, 2001. Available at http://www.nti.org/gsn/article/terrorists-could-exploit-hurricane-in-us-report-warns/, accessed June 22, 2012.

3. Wood S. DoD leaders report on hurricane response. American Forces Information Service News Articles, November 2005. Available at http://www.defenselink.mil/news/Nov2005/20051110_3310.html, accessed June 22, 2012. Director of Mobility Forces After Action Report, Joint Task Forces Katrina and Rita, October 18, 2005, Appendix B: Mobility Metrics Overview.

4. Henry J. Incomplete evacuation. Hurricane Katrina, Joint Center for Operational Analysis, Quarterly Bulletin, Volume VIII, Issue 2, June 2006.

5. Ibid., Coordination, Command, Control, and Communication, Lt Col Greg Gecowets, USAF.

6. JP 1 GL-9 commander's intent. A concise expression of the purpose of the operation and the desired end state. It may also include the commander's assessment of the adversary commander's intent and an assessment of where and how much risk is acceptable during the operation. Alberts and Hayes in their book Understanding Command use the term "command intent" to describe this concept since in their opinion there is no longer a single commander present in any reasonably large operation (page 88).

7. A separate but related question is whether cyber is a domain at all, or whether it is the electromagnetic spectrum that is domain, and cyber is a ship that enhances human's ability to exploit it. Similar to how air space is the domain, and aircraft allow for its exploitation. The air traffic corridors and other man made elements that allow for the exploitation of airspace as a technological system are not the domain.

8. Cite Rattray See also: Clark D. Characterizing cyberspace: Past, present and future. Working paper (2010). Available at http://web.mit.edu/ecir/pdf/clark-cyberspace.pdf, accessed June 22, 2012.

9. Clark, ibid, 1. The definition provided by the Joint Chiefs of Staff in the National Military Operations for Cyberspace is overly limited. The USAF should consider embedding the JCS definition within its doctrine.

10. TCP/IP is standardized by the International Organization of Standards (ISO) for the Open Systems Interconnection (OSI) model as the basis of Internet and other networking.

11. Lipson HF. Tracking and Tracing Cyber-Attacks: Technical Challenges and Global Policy Issues. CERT Special Report, 2002.

12. Leighton T. The net's real security problem. Scientific American, September 2006, 44.

13. Ibid.

14. Molyneux RE. *The Internet under the Hood: An Introduction to Network Technologies for Information Professionals*. Westport, CT: Libraries Unlimited; 2003. pp 85–86.

15. Ibid., 27.

16. Yannakogeorgos PA. *Essential Questions for Cyber Policy: Strategically Using Global Norms to Reduce the Cyber Attribution Problem.* Maxwell AFB, Alabama: Air University Press; 2011.

17. http://news.cnet.com/8301-31001_3-20077492-261/top-isps-agree-to-become-copyright-cops/, accessed June 22, 2012.

18. Ghernouti-Helie S. A national strategy for an effective cybersecurity approach and culture. 2010 International Conference on Availability Reliability and Security. IEEE Publications; 2010.

19. Gady FS. Africa's cyber WMD. Foreign Policy, March 24, 2010. Available at http://www.foreignpolicy.com/articles/2010/03/24/africas_cyber_wmd, accessed June 22, 2012.

20. From remarks by U.S. Secretary of State Hillary Clinton, January 21, 2010: "The spread of information networks is forming a new nervous system for our planet. . . . States, terrorists, and those who would act as their proxies must know that the United States will protect our networks. Those who disrupt the free flow of information in our society or any other pose a threat to our economy, our government, and our civil society. Countries or individuals that engage in

cyber attacks should face consequences and international condemnation. In an Internet-connected world, an attack on one nation's networks can be an attack on all. And by reinforcing that message, we can create norms of behavior among States and encourage respect for the global networked commons." Available at http://www.state.gov/secretary/rm/2010/01/135519.htm, accessed June 22, 2012.

21. John Markoff, SecurID Company Suffers a Breach of Data Security. New York Times, March 17, 2011. Available at http://www.nytimes.com/2011/03/18/technology/18secure.html?_r=1, accessed June 22, 2012.

22. I am grateful to Mr. Lynn Mattice for this observation.

23. The NCC was established by the National Security Telecommunications Advisory Committee (NSTAC) to serve as its operational arm. NSTAC has been advising presidents on cybersecurity since the 1980s.

24. NCC Operating Charter. Available at http://www.ncs.gov/ncc/nccoc/nccoc_background.html, accessed June 22, 2012.

25. U.S. Government Accountability Office. Report to the Subcommittee on Emerging Threats, Cybersecurity, and Science and Technology, Committee on Homeland Security, House of Representatives, Critical Infrastructure Protection: Further Efforts Needed to Integrate Planning for and Response to Disruptions on Converged Voice and Data Networks, June 2008, 11.

26. Ibid., 11.

27. Ibid., 12.

28. Ibid., 14.

29. U.S. Department of Homeland Security. Secretary Napolitano Opens New National Cybersecurity and Communications Integration Center, October 30, 2009. Available at http://www.dhs.gov/ynews/releases/pr_1256914923094.shtm, accessed June 22, 2012.

30. Rod Beckstrom. Letter of Resignation to DHS Secretary Janet Napolitano, March 5, 2009.

31. U.S. Department of Homeland Security. Memorandum of Agreement between the Department of Homeland Security and the Department of Defense Regarding Cybersecurity, October 13, 2010.

PART V

CONCLUSION

19

PREPAREDNESS, BLACK SWANS, AND SALIENT THEMES

Nancy D. Connell and Leonard A. Cole

As midnight approached on July 8, 1962, US scientists detonated a thermonuclear bomb 250 miles above the Pacific Ocean. The high-atmosphere megaton explosion lit the sky. Its flash was visible as far away as Kwajalein Island, a thousand miles to the southwest. The detonation also generated a massive electromagnetic pulse (EMP) that had surprise effects including the extinguishing of Hawaiian streetlights and the interruption of radio reception.[1] Subsequent thermonuclear tests confirmed the potential for EMP to take down communication systems.

Since then, the everyday use of electronic devices and telecommunication satellites has hugely expanded. Our increased dependency on these devices and systems has prompted concern among a few observers. "A massive EMP attack on the United States could produce almost unimaginable devastation by wiping out essential infrastructure," said security analyst James Carafano. "Communications would collapse, transportation would halt, and electrical power would disappear."[2]

Carafano's worries, however, have gained little traction. A spokesman for the Pentagon's Missile Defense Agency called the potential damage from an EMP attack "pretty theoretical."[3] Claims of devastation have been derided as part of the "EMP, scare-monger, worry-wart crowd."[4] Still, the EMP story was cited again by writer Warren Kozak on the 70th anniversary of the Japanese attack on Pearl Harbor.

Why December 7 and EMP in the same article? As Kozak noted, a Japanese attack against America's largest Pacific naval base seemed unthinkable. The US government could not conceive of, and did not defend against, such an assault.[5] But in 90 minutes Japanese planes had demolished most of America's Pacific fleet and its Hawaii-based aircraft.

December 7, 1941, was, in the parlance of this book, a *black swan*.[6] Could an EMP attack prove to be another one? Not, obviously, unless it happens. Kozak concluded his article with an acknowledgment that skeptics about EMP may be right. "But 70 years ago

Local Planning for Terror and Disaster: From Bioterrorism to Earthquakes, First Edition.
Edited by Leonard A. Cole and Nancy D. Connell.
© 2012 John Wiley & Sons, Inc. Published 2012 by John Wiley & Sons, Inc.

241

similar doubters believed Japan would never be so foolish as to take on the United States of America—until, of course, it did." Which presents a dilemma: how much, if at all, should the United States and other countries be preparing for unlikely events?

PREPARING FOR BLACK SWANS

Many chapters in this volume have been developed around case studies that were *black swans*. Because they had seemed improbable, in some instances unthinkable, they were not specifically part of anyone's preparedness plans. Anthrax spores leaking from letters, airline hijackers bent on crashing into buildings, tsunami waves far taller than anticipated, all resulted in tragedies made worse because they had seemed unfathomable. Even after these events were underway, neither professional responders nor the general public initially understood their gravity, which impaired taking early action to lessen their effects.

Yet some *black swans* described in this book were addressed effectively. Among the more successful was New Jersey's preparedness and response in 2011 to a surprise sequence of potential disasters. In quick succession the state experienced a mild earthquake, a severe hurricane, and rampant flooding. Recounted by Mark Merlin in Chapter 11, the overall response, based on an all-hazards approach to preparedness, went quite well.

Another example of a salutary response, though in an entirely different context, took place in Israel. Described by Shmuel C. Shapira and Limore Aharonson-Daniel in Chapter 9, the 2001 collapse of a wedding hall floor resulted in 23 deaths and hundreds of injured. The destruction was initially assumed to be terrorist related since it occurred at a time of many terrorist attacks against Israel. To have thought otherwise, that the event was *not* connected to terrorism, would make it somewhat of a *black swan*. Yet rapid and careful assessment ruled out the possibility of a deliberate attack, which enabled rescue operations to proceed with fewer impediments. (Israeli responses to terrorist events include time-consuming security precautions that might not apply to non-terrorist incidents.) The collapse was eventually confirmed as caused by faulty construction.

These examples suggest an aspect of proper preparedness that is relevant to *black swan* events. As suggested in Chapter 1, planning should include anticipation of the unexpected. Of course, exercises should focus on scenarios based on experience. But some should also incorporate surprise challenges that provide opportunities for responders to face the unexpected while under pressure and in real time.

When responders are well prepared for the expected, rescue and recovery are commonly timely and efficient. Israeli responses to terrorist attacks have repeatedly attested to this fact. In the United States, the value of solid preparedness was demonstrated by the response to the 1996 Olympic Park bombing in Atlanta, as recapitulated in Chapter 5. In these cases, actual events approximated the rehearsed scripts. It is when an event deviates from the expected, posing distinctive challenges, that innovation could be called for.

Prior to 9/11, airline protocols called for the crew and passengers of a hijacked plane to cooperate with the hijackers. No one had anticipated that the 9/11 hijackers, unlike others in the past, had massive homicidal intentions. A different mindset might have altered the outcome since it could have prompted intervention by passengers and crewmembers. Indeed, after passengers in the fourth jetliner learned that the other three had crashed into the Pentagon and the Twin Towers, they ignored the protocol and tried to overcome their captors. During the scuffle the plane crashed into a Pennsylvania field, though their brave action prevented it from reaching a population center.

A challenge to protocol can also arise during routine rescue of trauma victims. Should care be provided at the scene or should it be forestalled in favor of transportation to a hospital? The former approach is preferred in some jurisdictions and the latter, designated as "scoop and run," in others. The rationale for on-scene care accords with the notion that early treatment enhances the chance for a favorable patient outcome. Conversely, scoop and run means quicker arrival at a hospital where advanced technology and medical services are available.

Doubtless, a decision should depend on a variety of considerations including the severity and nature of injury, skill level of the responders, and proximity of a hospital. But even when accounting for these variables, a protocol may appear unduly rigid. From an article on resuscitation of hemorrhagic shock:

> When evacuation time is shorter than one hour (usually urban trauma) immediate evacuation to a surgical facility is indicated after airway and breathing have been secured ("scoop and run")... When expected evacuation time exceeds one hour an intravenous line is introduced and fluid treatment started before evacuation.[7]

In aspiring to precision, the protocol instead suggests undue rigidity. Evacuation times cannot always be predicted, so how should a responder act in the face of uncertainty? How literal is the one-hour admonition? Some responders, while believing that an on-scene patient would benefit from early fluid treatment, might nonetheless hesitate to act. They might not feel secure in their own judgment or might fear disciplinary action for violating a protocol. Bypassing existing protocol is not a trivial matter. When such action may be justified is not always clear, but acknowledging that it may be warranted under some conditions is an essential first step.

Besides deviation from standard procedures, reaction to a *black swan* also has an emotional component. As Nancy Green, an executive with Aon Risk Solutions, astutely observed: "The key to addressing a *black swan* is not just mounting an effective response; it is mounting that response while simultaneously dealing with the psychological impact of being shocked by an inconceivable event of staggering proportions."[8]

Accordingly, hands-on rehearsal for an improbable event can impart confidence—psychological readiness—that could be essential to an effective response. While a surprise scenario might call for innovative action, this does not imply an ungoverned free for all. Rather it requires a leadership that recognizes when a situation is extraordinary and when extraordinary responses are necessary. The biggest obstacle in such a situation is a bureaucratic mindset that rejects any and all actions that have not been preapproved.

SALIENT THEMES AND OBSERVATIONS

The chapters in this volume have explored numerous aspects of human disasters and catastrophes, whether natural (earthquakes, tsunamis, floods), accidental (industrial leaks, structural failure), or terrorist-related (biological attack). Collectively they have produced several unifying themes as well as some discrepancies. Chapter authors considered and analyzed the roles of both central and ancillary responders and explored the behavioral effects of trauma. In addition to traditional responders such as medics and law enforcement personnel, new classes of players and novel aspects of emergency medicine have been brought into focus. A next step would be to formalize the new responder groups and management approaches and incorporate their potential contributions in planning and exercises. Following are several major themes and observations that have threaded through the chapters.

Emergency Communication (Technologies)

From beginning to end in any mass casualty incident (MCI), communications are critical to an effective response. Many of the authors observed that communications were an essential part of the response toolkit. Communication problems were a focus of concern in hearings and post-event analyses of the emergency response to the 9/11 attacks, the London bombings in 2005, and the Fukushima radiation leaks in 2011. The specific technologies—cell, satellite, web-based, land-based—frequently failed or otherwise proved to be inadequate (Chapters 1–3).

Authorities may first be notified that an event has taken place by survivors or bystanders. Civilian cell phones should therefore be able to link effectively with emergency response authorities. But failures in interoperability and impediments to speed and accuracy were common in many disaster situations reviewed here. In fact, cell phones did not operate at all in the London underground.

Mechanisms for information exchange among responders, government, media, and the public are essential and should be tested frequently. Israel provides a good model. While interconnectedness between police, fire, hazmat, and command posts does not always re-main active there, the networks are continually tested during intervals between emergencies (Chapter 9).

Communication (Risk)

A second key aspect of communications is that of risk communication. Several authors were critical of how the public was informed about an event or about continuing danger as the event unfolded or in its aftermath. Of special concern were the methods of communication between elected and appointed officials and between all officials and the public. For exam-ple, a bioterrorism attack bears features that are distinctive from more "traditional" methods of terrorism such as bombings, shootings, and other forms of explosives. The scientific and public health issues surrounding bioterrorism can be complicated and sometimes unclear to officials themselves.

Several chapters included discussions on the difficulties of informing the public with appropriate detail, precision, and instruction. The US anthrax attacks of 2001, for example, offer a chilling reminder of the urgency of correct information (Chapter 2); some deaths and infections could have been avoided by prompt communication of correct information.

Similarly, as discussed in Chapter 3, incorrect or inadequate information communication led to long-term health risks in the Chernobyl and Bhopal accidents. In the more recent Fukushima Dai-ichi disaster, the impact of radioactivity on health was not properly com-municated in the early stages of the disaster because of ill-defined roles and poor lines of communication between the utility and local and national authorities.

During and after Hurricane Katrina, reliable information regarding the status of victims and other vital matters was often more available through the media than from state and federal agencies (Chapters 4 and 16). In Chapter 3, Becker described seven principles for effective risk communication that are particularly salient: Messages to the public must be

(1) proactive, (2) developed and rehearsed during planning stages, (3) provided by a variety of channels, (4) practical and actionable, (5) clear and easy to understand, (6) credible, and (7) consistent. Overall, plans and exercises for risk communication to the public during an MCI should be addressed with no less urgency than plans and exercises for medical responses.

Adaptation to Disaster Mode

As a disaster unfolds, the human components—victims, bystanders, responders, managers—must adapt as conditions change. Many chapters describe shifting circumstances and the need for people to make accommodations under sometimes shocking conditions. In Chapter 9, on hospital management, Shapira and Aharonson-Daniel discuss the "rapid transformation of a slow and complex system to one that addresses a sudden influx of patients in urgent need of care."

Since the introduction of the National Incident Management System (NIMS) in 2003, the United States has solved some interoperability and response problems. But several communication issues persist and Newman and Clarke in Chapter 16 urge movement "beyond the static NIMS model." We must find ways for critical information to move efficiently from the ground or site up through the chain of command and back down again.

Leadership should enhance the flow of information and promote rapid adaptation to crisis conditions. Yet most management systems seem unwilling to abandon the concept of "top-down command," and few individuals in leadership positions have both the talent and authorization to create "linkages beyond the confines of . . . bureaucratic entities" (Chapter 9). Rectification of common pitfalls is crucial to maintaining appropriate response capabilities. These pitfalls include distraction, identifying a threshold for withholding medical treatment, keeping flexibility and an open mind under pressure, and learning to deal with bureaucracies and bottlenecks.

Fluidity of Roles

The professional identities of typical responder groups in a disaster situation are familiar: EMS, physicians, nurses, police, fire, hazmat. But this book identifies additional actual or potential responders who could be key assets during a terrorist or disaster event. They include bystanders and survivors (Chapter 13), faith-based and other citizen volunteers (Chapter 14), and laboratorians (Chapter 15). In Chapter 6, Glotzer argues that dentists are an especially valuable resource and should be included in planning, training, and exercises in response efforts.

Inclusion of certain novel groups could also solve some problems now encountered at disaster sites. Bystanders and survivors, although lacking appropriate skills, may insist on "helping." In Chapter 13, Adini points out that in traumatic situations they may not perceive that they are hindering response activities. But harnessing their eagerness to help could transform them from an obstructive to a constructive presence.

Training opportunities for on-site management of would-be bystanders are unlikely to draw much interest. Still, incident commanders should be aware of how they could be usefully engaged in roles ranging from transporting and comforting the injured to directing traffic away from the site.

Several authors suggested that novel responder groups be formally recognized, and that planning exercises incorporate these less traditional players in the response network.

Fostering Resilience

How can we build disaster-resilient communities? Despite differences among societies in size, demographics, and other characteristics, some basics about preparedness and response are common to all. After a major earthquake in Italy in 2009, a large number of dialysis patients were moved to distant locations without a delay in treatment. Similarly, after flooding in New Jersey in 2011, many elderly and infirm residents were expeditiously transported to safe shelters. Success in both instances was due to careful and effective planning (Chapter 11).

Israel's MCI planning is regarded by many as an effective model that also promotes resilience (especially in the face of terrorist threats) (Chapters 1 and 10). Although the United States is a much larger country, many of Israel's approaches are applicable, especially since responses here are commonly regarded as local for the first 48 hours. Several chapters refer to novel solutions found in Israeli approaches to disaster management.

Implicit in many narratives is the fact that resilience depends heavily on an effective response capability. This requires three primary elements: people, resources, and planning. Established volunteer organizations such as the Red Cross and faith-based groups are important to the process. Civic-minded volunteers, not necessarily trained for disaster response, should also be incorporated into planning models, to coordinate blood donations and contributions of money, supplies, and comfort (Chapter 14). Essential contributions toward resilience also come from mental health professionals, and their integration into disaster response is well described in Chapter 8.

Planning–Training–Exercising

The combination of planning, training, and exercising is the premier trio required to maintain a strong and prepared response system. Many chapters have stressed exercises as fundamental to the realization of disaster planning, largely through workshops, tabletops, or other drill formats. Interestingly, there was one dissenting view: Newman and Clarke in Chapter 16 advocate "planning and partnering" in terrorism preparedness, maintaining that an integrated and operational crime prevention network—which already exists in major US cities—is fully capable of dealing with terrorist crimes. Sinai, in Chapter 17, gives an in-depth description of a real-world case using integrated investigative cooperation across state and international borders.

Despite the prevailing view that constant reworking of plans, accompanied by training and exercises, is essential to effective disaster response, there are a number of failings in the contemporary planning and training approaches. For example, training frequently is wanting in several germane areas including clarity of chain of command, elementary medical literacy among nonmedical responders (fire, police), and under or over triage (mismatching the number and needs and of patients with the capabilities of a receiving hospital) (Chapter 11).

Furthermore, current planning often does not adequately address human behavioral issues (Chapters 8 and 12). There is rarely any interaction with the public in disaster planning and this leads to gaps between what emergency responders expect or want people to do and what they actually can do. Constant updating and modification of approaches to planning and training is necessary to build a disaster-resilient community.

A FINAL OBSERVATION ABOUT PREPAREDNESS

In the months after 9/11 and the anthrax attacks, polls indicated that terrorism had become the US public's top concern. A decade later, less than 1% of respondents deemed it the most important problem facing the country.[9] This diminished sense of urgency troubles security and other officials, who believe the threat remains unabated.

Mirroring the downturn in public concern, financing for emergency preparedness, especially at hospitals, has been shrinking. As described in Chapter 5, Arthur Kellermann, former director of emergency medicine at Emory University, has despaired in particular about the deterioration of emergency medical response capabilities.[10]

In the wake of 2001, money from the US Departments of Homeland Security and Health and Human Services at first streamed into the states for emergency preparedness. Hospitals were major beneficiaries, though by 2006 federal support had begun to decline. In New Jersey, for example, hospitals that had received $150,000 or more in 2005 for protective equipment, drills, and educational programs saw financing fall to as little as $20,000 just 2 years later. As a result, the state's emergency readiness was weakened.[11]

Many hospitals conduct one or two drills a year for terrorism and disaster events. But the quality of these exercises varies. Some conduct full-scale drills with dozens of participants whose performance is carefully evaluated. But according to a New Jersey state health official, drills at other hospitals may involve only "three people sitting in a room, sharing information with each other, with no follow-up." The problem is compounded as hospital chiefs scramble to find scarce dollars for daily upkeep.

Fire Drills: A Model for Disaster Exercises

A solution would be to mandate standards for emergency preparedness much as is done for fire protection. By law, New Jersey hospitals must conduct at least four fire drills a year that include personnel from every shift. The costs of these exercises, including the maintenance of smoke alarms, extinguishers, and other equipment can be as high as $50,000 a year per hospital. No one questions the need or expense for this kind of protection even though almost none of New Jersey's hospitals ever experienced a major fire. Yet many of these institutions were involved in treating evacuees from Lower Manhattan on September 11, 2001.

Jersey City Medical Center treated 175 victims who were ferried across the Hudson River, and the University Hospital in Newark saw about 50. A month later, anthrax spores were found to have leaked from letters that were processed at a postal distribution center in Hamilton, near Trenton. Three infected victims were treated at nearby Robert Wood Johnson Hospital, which had to test more than 1300 others who had been exposed to the deadly bacteria. An additional 400 suspected anthrax victims were seen at St. Francis Hospital in Trenton.

A statewide mock biological attack in 2005 demonstrated huge differences among hospitals in their ability to cope with a large-scale emergency. The exercise underscored the need for mandated standards that would include three imperatives. First, every hospital should be able to increase normal capacity by at least 20% in case of a sudden influx of victims. Second, all hospitals should have a staffer devoted to emergency preparedness.

Third, rehearsals for biological, chemical, and other disaster events should be conducted according to uniform procedures.

CONCLUSION

Many issues about hospital readiness pertain as well to other institutions. Fire drills are also conducted routinely in schools, government offices, and commercial buildings. These structures may contain hundreds or thousands of occupants. Their large numbers compound the risk of congestion or even chaos in the event of a terrorist or disaster event. Just as periodic safety and evacuation drills for fire are deemed necessary in these buildings, so should exercises for terrorism and disaster be part of their preparedness efforts.

The commentaries in this volume have revealed many strengths as well as weaknesses in current local preparedness. They also demonstrate the extraordinary breadth and variety of fields dedicated to caring for fellow citizens in the event of an MCI. That so many people have committed themselves to helping their neighbors is as important and heartening as any other message in this book.

NOTES

1. Vittitoe CN. Did high-altitude EMP cause the Hawaiian streetlight incident? System Design and Assessment Notes. Sandia National Laboratories. Albuquerque, NM, June 1989.
2. Carafano JJ. End of the world, for real. The Washington Examiner, August 15, 2010. Available at http://washingtonexaminer.com/article/36314, accessed June 22, 2012.
3. Broad WJ. Among Gingrich's passions, a doomsday vision. The New York Times, December 12, 2011; A-1, A-4.
4. Shachtman N. Interview on National Public Radio, November 23, 2011. Available at http://www.npr.org/2011/11/23/142723604/is-an-electromagnetic-pulse-attack-a-threat, accessed June 22, 2012.
5. Kozak W. Pearl Harbor, Iran and North Korea. The Wall Street Journal, December 7, 2011; A-17.
6. See the section in Chapter 1 titled "The Known, the Unknown, and the Black Swan," and also Taleb NN. *The Black Swan: The Impact of the Highly Improbable*. 2nd edn. New York: Random House; 2010.
7. Krausz MM. Initial resuscitation of hemorrhagic shock. World J Emerg Surg 2006;1:14.
8. Green N. Keys to success in managing a black swan event. Available at http://www.aon.com/attachments/risk-services/Manage_Black_Swan_Even_Whitepaper_31811.pdf, accessed June 22, 2012.
9. Bowman K, Rugg A. Americans and the terrorism threat 10 years after 9/11. The American, August 31, 2011. Available at http://www.aei.org/article/americans-and-the-terrorism-threat-10-years-after-911/, accessed June 22, 2012.
10. Kellermann AL. Crisis in the emergency department. N Engl J Med 2006;355:1300–1303.
11. Much of the remaining narrative is excerpted from Cole LA, Asleep in the E.R. The New York Times, June 10, 2007.

INDEX

Local Planning for Terror and Disaster: From Bioterrorism to Earthquakes, First Edition.
Edited by Leonard A. Cole and Nancy D. Connell.
© 2012 John Wiley & Sons, Inc. Published 2012 by John Wiley & Sons, Inc.